Protein Purification

Protein Purification

Edited by:

Bhupendra Pushkar

Delve PUBLISHING

www.delvepublishing.com

Protein Purification

Bhupendra Pushkar

Delve Publishing
224 Shoreacres Road
Burlington, ON L7L 2H2
Canada
www.delvepublishing.com
Email: orders@arclereducation.com

ISBN: 978-1-77407-702-3 (Hardcover)

ABOUT THE EDITOR

Dr. Bhupendra Pushkar was born in Mumbai, India. He persuaded his B.Sc. in Microbiology in 2000. He has obtained his M.Sc. degree in Biotechnology in 2002 from University of Mumbai. He pursued his further studies and completed his Doctorate degree in 2010. His area of research broadly includes environmental pollution remediation using biological science. He is currently working on heavy metal bioremediation, PAH bioremediation, plastic degradation, plant tissue culture etc. He is actively involved in the bioremediation research work and trying to develop system for river water cleaning. Since 2008 he is working as assistant professor at University of Mumbai, India. He has published various papers in national as well as international journals and currently guiding three research students. He is Co-Editor in Chief of a journal and reviewer in many journals.

TABLE OF CONTENTS

LIST OF FIGURES

PREFACE

Protein purification can be considered as an essential step in many research laboratories. It is important to know the efficiency and limitations of the purification protocols. There are several protein purification techniques than are successful at the laboratory scale and even at the industrial scale. The protein purification protocols must be flawless as they are the determinants of the proteins obtained may be used in the food and health sectors. Protein chemistry and purification tools are essential in current biological studies and research. Due to developments in the heterologous expression of the proteins via recombinant DNA technology, there is an increased demand for advanced protein purification protocols. The main aim of all the advanced and conventional techniques is to purify the proteins and characterize them. The protein electrophoresis can be used in clinical diagnosis and life sciences. This book can be used by the readers who are willing to improve the skills inanalytical science. This book may be used in understanding the role of protein purification protocols in protein therapeutics. This book can serve as a timely reference for those who are interested in analytical protein chemistry. Mass spectroscopy with the traditional electrophoresis can yield better results thanconventional techniques. In pharmaceutical industries, the recombinant proteins are becoming the workhorses and there is a demand for advanced techniques. Proteomics and genomics have opened several opportunities for researchers all over the world and these areas demand the highest accuracy in the protein purification methods.

This book can serve as a handbook for undergraduate, postgraduate and research students who are interested to study proteins. This book deals with both electrophoresis and chromatography techniques. Mass spectroscopy in protein purification protocols was also discussed. The latest equipment and standards were explained in this book. Advanced chromatographic techniques such as High-Performance Liquid Chromatography have been discussed in this book. Nanotechnology in the protein purification protocols was discussed in detail.

We hope that this book can satisfy the needs of students who are involved in study and research experiments aboutprotein dynamics and structure. Most of the technical knowledge concerning the protein purification techniques was extracted from contemporary resources to enlighten the students.

INTRODUCTION

Most of the proteins undergo folding and so they get a specific shape. The proteins will be in three-dimensional conformation. It is important to separate a protein from a complex mixture and it is to be ensured that the native biological function is maintained. We can maintain the native biological function by keeping critical parameters steady and they are ionic strength, temperature and pH. The selection of protein sources can be from insect and mammalian cells, yeast cells, *E. coli* cells and other natural resources. The stability and solubility of the proteins must be maintained. This can be possible by using the crude protein extracts and stabilizing the parameters needed to keep the proteins at their native state. The proteins must be subjected to assays and there are specific. The quantity and quality of the proteins to be purified is important. The protein purification strategies can be based on the surface charges present on the proteins and molecular size.

In the last 30 years, there are several developments in the polyacrylamide gel electrophoretic methods that resolved several problems of protein purification. 1D-PAGE was classified as a simple technique relatively. There are several variants of PAGE that can help us purify proteins at various levels. HPLC can be classified as an important technique in the analysis and purification of different proteins. With HPLC, we can expect high rates of reproducibility and manipulations can be easy. We can expect high rates of recoveries. Reversed-Phase chromatography can be used to separate the peptides. The agarose and polyacrylamide gels can be used in the electrophoretic techniques. We can purify the nucleic acids as these gels are porous and we can use that property in the separation process.

Nanotechnology can be integrated with the electrophoresis process and chromatographic techniques. The nanoparticles can be diverse functions that are important in numerous assays and life science applications. We can use nanotechnology-based assays in drug delivery studies. We can use luminescence and fluorescence agents along with the nanoparticles so that it can be studied with more depth.

1

Introduction to Protein Purification

CONTENTS

1.1 INTRODUCTION

The purification of proteins may involve several processes that are used to isolate one or more proteins from a mixture. This mixture may consists organism, tissues or cells. The protein purification process is important in characterizing the structure, function and interactions with other proteins or other compounds. In the purification process, there will be a separation of both protein and non-protein components in the mixture. This will help us to isolate the desired protein from all non-essential proteins. The separation of the desired protein can be a tedious process in the entire process of protein purification. Most of the steps involved in the process of separation can use some important properties such as biological activity, binding affinity, physicochemical properties and protein size. The protein obtained after the purification process is called as 'protein isolate'. The protein purification process may follow two modes. The first mode is the preparative mode and the second mode is called an analytical mode. In the preparative mode of purification, there will be the production of massive quantities of purified proteins which can be subjected to further processing. We can produce insulin, soy protein isolate and lactase using this purification mode. Many commercial products such as biopharmaceuticals, nutritional proteins and enzymes can be produced by this preparation mode. In the purification process, most of the bi-products will be removed to ensure the patient's health.

In the case of analytical mode of purification, there will be a minute amount of protein and the protein isolate will be useful for analytical purposes. In this mode, there will be identification and quantification. This purification mode can help us to understand the structure of protein, function and post-translational modifications. Some of the proteins produced with this mode are urease and pepsin. These proteins can be crystallized after purification. There are some instances where protein may not be secreted or released from the organisms. The immediate solution for this is the cell disruption and forced release of proteins. The fragility of both the cells and proteins can be important in the purification process. There are several methods that can help us to force release the proteins in the surrounding solution. The cells can be permeabilized for detergents and enzymes such as Triton X-100 and lysozyme. The homogenization of the cells can be performed by bead mills and the process is called grinding. We can use French press to homogenize the cells with high pressure. Sonification can be used for this purpose. We can do repeated thawing and freezing so that the proteins will ooze out from the cells. The cell debris can be extracted with the aid of centrifugation. Some of the soluble proteins can be present in the supernatant. Most of the proteases are released in the process of cell lysis. These proteases can digest the proteins which are present in the solution. In case the desired protein is susceptible to proteolysis, it is important to make the process faster. It is important to maintain cool temperatures so that the digestion will be slowed down. We can use protease inhibitors with the lysis buffer before cell disruption. It is important to add deoxyribonuclease or DNAse so that there will be reduced viscosity in the lysate. The viscosity can be due to higher DNA concentrations. In the case of bulk purification of proteins, the primary step is to isolate the proteins. The isolation of proteins can be done via the precipitation process. The precipitation process can be done with ammonium sulfate. We can increase the concentration of ammonium sulfate and extract different fractions of the proteins. We can remove the excess amounts of ammonium sulfate via dialysis. Whilst the precipitation process, the hydrophobic groups will be exposed and there will be the formation of hydrophobic components. The precipitate can be observed with the naked eye. The main advantage of this method is that it is cheaper. The proteins which will be purified first are classified as water-soluble proteins. The proteins which are present in the membranes are called integral membrane proteins. These proteins can be isolated only if the cells are disrupted. After the cell disruption process, the proteins can be isolated. In some cases, it is easier to isolate mitochondria than proteins present in the membranes of the mitochondria. Sodium dodecyl sulfate or SDS can be added to the cells so that the cell membranes will be disrupted. The SDS molecule can be used in the denaturation process. We can use detergents like CHAPS or triton X-100 so that the native conformation of the proteins will be intact even after purification.

Ultracentrifugation (Figure 1.1) is another important technique which is widely used in the protein purification process. In the centrifugation process, the centrifugal force is used to separate the particles from mixtures according to their densities. When a bottle or a tube with cells rotate at high speeds, the particles get inertia and it is proportional to the mass of the particles. Due to spinning, the cells present in the sample can be separated based on their respective masses. If suspensions are centrifuged, then there will be the formation of pellets at the bottom of the tube. The non-compact materials can be present in the supernatant. We can extract the supernatant from the pellet. The centrifugation rate can be defined by the angular acceleration and it is measured in 'g'. In the case of sucrose gradient centrifugation, the sugar concentration is high at the top and low at the bottom. The protein sample is added to the sucrose gradient solution and is allowed to centrifugate. We can use ultracentrifugation in this case. Due to this, the heavy particles can stay at the bottom and the lighter materials on the top. With this technique, we can counteract the centrifugal forces and the motion of the particles will be in close proportion with time. The samples which are separated by this technique can be called as rate zonal centrifugations.

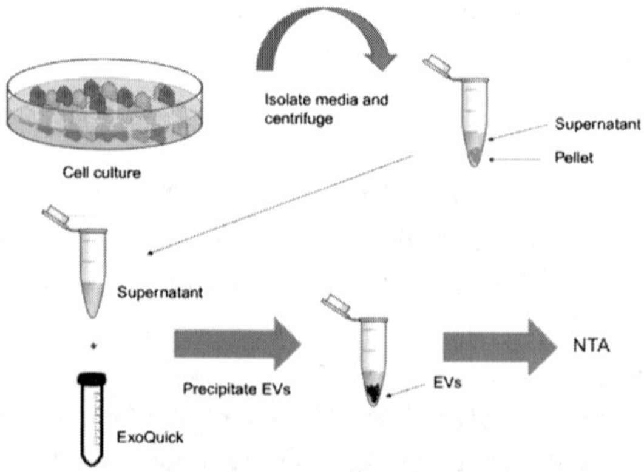

Figure 1.1. Ultracentrifugation.

Source: https://www.sciencedirect.com/topics/chemistry/ultracentrifugation

The source or the sample material is important in any purification process. A protein can be homogeneously distributed all over the organism in most of the cases. Some tissues or organs can have high concentrations than others. We can use the tissues which can have high protein concentrations and this will lower the need to isolating large tissues. In case there is less protein availability, researchers use rDNA technology. Through this technology, it is easy to produce cells can secrete proteins. This can be called an expression system. In the recombinant expression systems, the proteins will be tagged. Due to the tagging process, the purification of proteins will be easier. Strep-tag and His-tag can be classified as examples of this process. The analytical mode of purification uses 3 main properties. The proteins can be purified based on isoelectric points. This can be done with the pH graded gel. In some cases, we can use the ion exchange column. The proteins can be subjected to separation and it will be based on molecular weight and size. We can use SDS-PAGE and size exclusion chromatography to separate the proteins according to their sizes. 2D PAGE can be used to purify the proteins. Researchers are using peptide mass fingerprinting to know the protein identity. We can use high-performance liquid chromatography or HPLC in which the proteins can be separated according to their polarity. In some places, reversed-phase chromatography is being used for this purpose. In the entire process of protein purification, there will be a need for more chromatographic steps. In general, chromatography involves migration of protein solution via a column which is previously packed with other materials.

It was established that different proteins react with column materials differently. Due to these reactions, there will be a difference in times of protein migration in the column. There are many chromatographic techniques available. The size exclusion chromatography can be used to separate the proteins present in the solution. It can be done with the help of porous gels. The molecules with less molecular weight and size can traverse into a large porous matrix. The proteins according to their size need an eluent which must be prepared carefully. We can collect the proteins at the end of the column gel with the eluent buffer. The hydrophobic interaction chromatography is based on the charge. The media used in this technique is amphiphilic in nature. It has both hydrophilic and hydrophobic regions. With this technique, we can separate proteins according to their hydrophobicity on the surface. In this technique, the proteins are subjected to a limited rate of denaturation and so we can preserve the native nature of the protein. In the case of ion-exchange chromatography, the proteins are separated according to their ionic charge. We can select the column by taking into consideration the strength and type of charge. The anion exchange resins can hold a positive charge and so the anions can be retained and in case of cation exchange resins, the cations will be retained. Before the separation, the buffer is added and passed via column. The solute molecules will interact with the buffer ions and so there will be competition for sites. The retention time will be based on the charge strength. The compounds which are having weak charge can be eluted fast than the compounds with stronger charges. The separation of the proteins can be regulated by temperature, the concentration of buffer, type of buffer, pH and mechanism of separation. This type of chromatography can be classified as a powerful tool in the protein purification process. This technique can be frequently used in both preparative and analytical separations. Free-flow electrophoresis or FFE can be an important technique in the protein purification process. It is carrier-free electrophoresis. The protein separation can be done using the orthogonal electric field. It can be used in the preparative mode of separation. We can use a laminar buffer in this technique. The pH gradient can be achieved by ampholytes. With this technique, we can separate protein isoforms.

Affinity chromatography (Figure 1.2) uses molecular conformation in protein separation. It uses resins. The resins will have ligands and they are linked on the surface. In this technique, we can observe the 'lock and key' method. It is highly specific in nature. Lectin affinity chromatography can be used to purify glycoproteins. The immunoaffinity chromatography can be used to purify proteins and it uses antibody-antigen interactions. In this process, there will be immobilization of proteins preferably on a solid surface. The proteins will bind to the target. The desired protein can be eluted by altering the salinity and pH of the solution. The immobilized ligand can act as an antibody or a protein. We can use this technique for proteins that were isolated from natural sources.

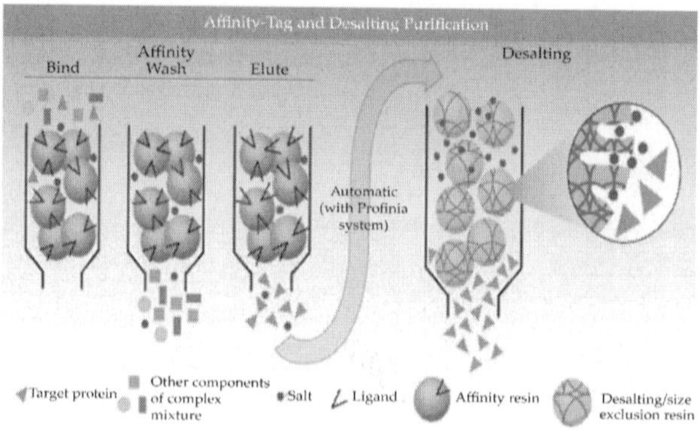

Figure 1.2. Affinity Chromatography.

Source: https://www.bio-rad.com/en-us/applications-technologies/introduction-affinity-chromatography?ID=MWHAVG4VY

In the process of tagging a protein, we can use the antigen peptide. After tagging, we can purify the protein from the column or resin which was previously coated with antibody. We can call it as immunoprecipitation. Through this technique, we can achieve more accuracy. The proteins which are previously tagged can be separated from the protein mixture. High-performance liquid chromatography involves the application of high pressure to move the solutes via columns at faster rates. Through this technique, the rate of diffusion will be reduced and so there will be an enhanced resolution. There is another variant called 'Reverse Phase HPLC' in which we use hydrophobic columns. Most of the proteins obtained from HPLC can be denatured.

After protein purification, the proteins obtained must be concentrated. The purified proteins can be subjected to lyophilization. In common terms, this process is called 'drying'. This process can be done with the proteins purified by HPLC. This technique removes all the compounds that are volatile in nature and the pure form of protein remains.

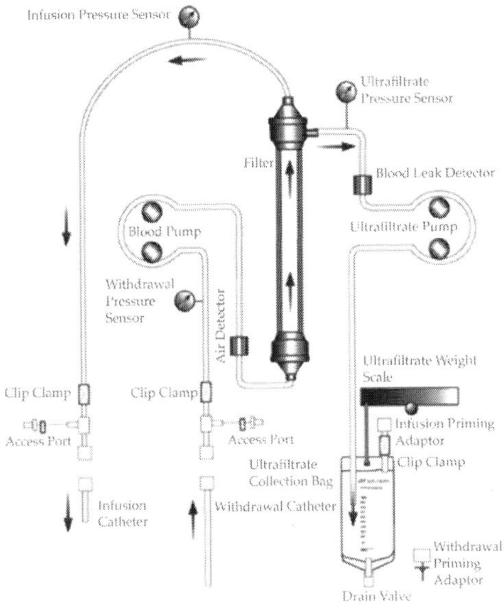

Figure 1.3. Ultrafiltration.

Source: https://www.sciencedirect.com/topics/medicine-and-dentistry/ultrafiltration

In the case of ultrafiltration (Figure 1.3), the protein solution is concentrated with the aid of selectively permeable membranes. The membranes used in this process will allow water along with other compounds and not the proteins. In this method, force is applied to the solution by centrifugation and gas pressure. The purification process can be evaluated by SDS-PAGE. With this method, we can have a vague measure of the proteins present in a mixture. We cannot demarcate proteins that share the same molecular weight. In case, the desired protein has unique spectroscopic features or specific enzymatic activity, we can use the these properties in the separation process.

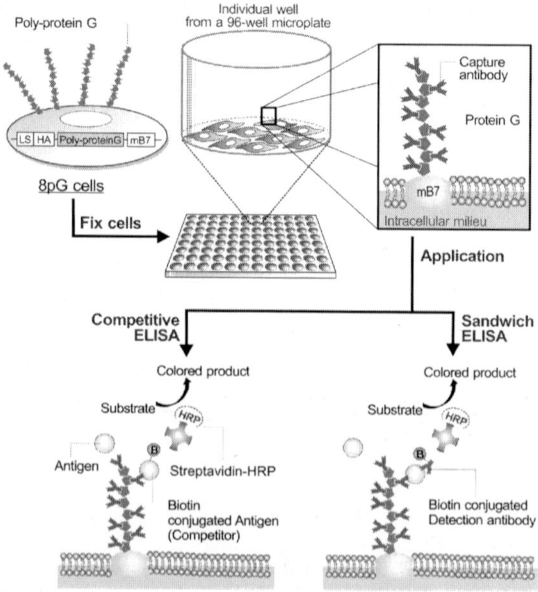

Figure 1.4. ELISA.

Source: https://www.nature.com/articles/s41598-018-36192-8

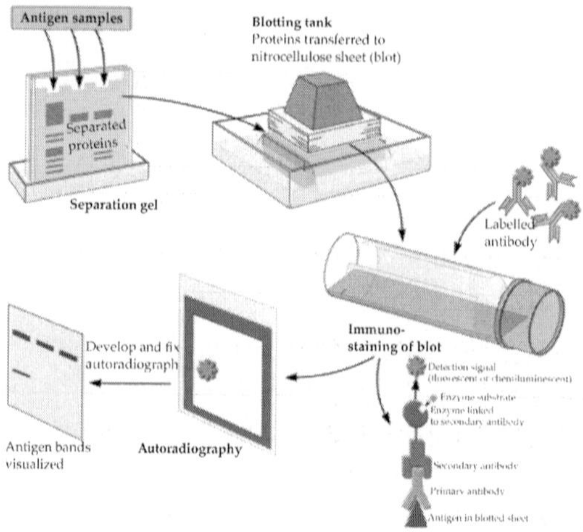

Figure 1.5. Western Blotting.

Source: https://www.sciencedirect.com/topics/medicine-and-dentistry/western-blotting

We can use ELISA (Figure 1.4) or western blotting (Figure 1.5) to detect the proteins with the aid of antibodies if available. There are some proteins that can work as receptors and we can use radioactive ligand in this case. In the evaluation process, the purified protein can be compared with the total protein. This can be done with the Bradford protein assay. The absorbance peak can be seen at 280nm. There are some reagents that can intervene in the protein purification process. Surface Plasmon Resonance or SPR (Figure 1.6) can be used in evaluating the protein purification process. This technique can detect the binding process. It can identify the molecules which are not labeled on the chip. Gel electrophoresis can be classified as an important lab technique which can be used in the analytical and preparative method. The main principle of electrophoresis is the migration of the charged ions. The desired proteins will be denatured if SDS is used.

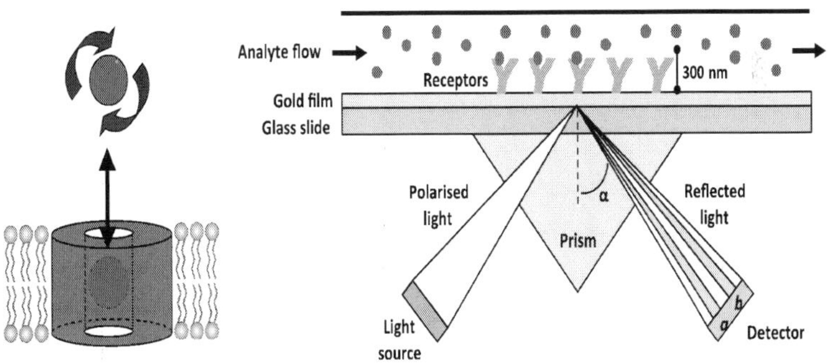

Figure 1.6. Surface Plasmon Resonance.

Source: https://www.sciencedirect.com/science/article/pii/S0005273613001466#f0005

The genome sequencing of several species has completed and this has generated demand to study molecular biology (Shendre & Lieberman, 2012). Proteomics is classified as an important tool in molecular biology. There is a need for the development of purification methods so that the protein can be used for various purposes (Carissa et al, 2012). The purified protein must be contaminant-free and it should not have isoforms produced due to translational modifications (Kalyanpur, 2002). During World War II, Edwin Joseph Cohn developed isolation and purification techniques (Cohn & Edsall, 1943). In the protein purification process, chromatography can be considered as an important tool (Lucy CA, 2003). The affinity chromatography technique was described by Emil Starkenstein (Starkensein, 1910). He separated alpha-amylase by using starch as an immobilized substrate. Affinity chromatography is coined by Meir Wilchek, Chris Anfinsen and Pedro Cuatrecasas (Cuatrecasas, 2004). In the protein purification process, the downstream processing needed certain factors so that the entire purification process can be cost-effective. They are cost efficiency, resolving power and higher yields. These needs have pushed us to improve the process tools and upgrade the strategies. Most of the proteins may not be purified with a single step (Butler *et al*, 2007& Wilken *et al,* 2011). Two-phase partition systems and fractional precipitation methods were used in low- resolution protein purification (Rosa PA, 1999). The type of chromatography to be used should be selected before the start of the purification process to purify the desired protein (Chon *et al,* 2004). This technique is common in downstream processing (Freitag *et al*, 1995). Affinity chromatography is better than most of the chromatographic techniques (Roque *et al,*2007). As the purification demands increased, there was a need for producing selective ligands (Labrou *et al,* 2003 & Clonis YD, 2006). It is beneficial to use synthetic ligands. Due to sophisticated techniques such as NMR, X-ray crystallography and bioinformatics, the selection of high- affinity synthetic ligands become easier and faster. We can use three methods to produce a ligand. They are the rational method, combinatorial method and structure-guided combinatorial method. In the biotechnology industries, the void maintained between the upstream and downstream processing can be considered as a major problem (Gottschalk, 2002). There is a shortage in the protein purification capacity in biotechnology industries. It became necessary that the entire protein purification process should be made economical. The basic information needed for the production of both safe and effective products was given by the International Society of Pharmaceutical Engineers or ISPE (ISPE, 2009) and International Organization for Standardization or ISO (ISO 14644-1, 1999). The proteins being used in the pharmaceutical industries need high purity levels, so the downstream processing has become expensive (Rasmussen *et al,* 2012 & Kelley B, 2009). The rDNA technology affects the protein purification process. There are many protein expression systems that can be present in the microbes, plants and animals but not limited to (Huang CJ *et al,* & Chen R, 2013). The production of fusion proteins became popular (Young CL, 2013). Most of the purified proteins were used as therapeutics (Walsh G, 2010).

1.2 PURIFICATION STRATEGIES

There are many approaches through which the proteins get purified. At every step, there will be a minimal protein loss. It is important to reduce the number of purification steps so that the loss of protein will be lowered considerably. The important aim of the protein purification process is that the process must be economical and should be less time- consuming. It is important to set the objectives needed with purity. The biological activity and quantity of the purified protein must be kept in mind. Most of the information regarding the contaminants and target protein properties can be important in the protein purification process. It is important to devise a reliable analytic assay which is faster. The extraction procedures and sample preparation must be studied prior to the start of the purification process. The purification of proteins can be dealt with three important phases that are named as capture, intermediate purification and polishing. In the capture step, there are some other steps called isolation, concentration and the last step will be stabilization of the target proteins. In the polishing stage, the impurities will be removed and some closely related substances will be removed.

The first step in the purification process is setting objectives for purity and quantity. We need to define the purity requirements. The *in vivo* studies and therapeutic applications demand extremely high purity which can greater than 99%. In the case of X-ray crystallography studies, we need high purity which can range from 95 to 99%. If we need to use the purified protein as an antigen, then the required purity can be moderate and that means 95% purity. Some protein properties may affect these strategies. The hydrophobicity may determine the medium to be used in the chromatography technique. The post-translational modifications can define the groups needed in the affinity medium. The biospecific affinity determines the ligand to be used in the affinity medium. The charge properties determine the ion exchange conditions that are to be used. The molecular weight of the purified protein will define the gel filtration media. The redox sensitivity of the protein give us an idea about the reducing agents which are to be used. The metal ion sensitivity can be helpful to decide which buffers are to be used. The protein sensitivity to proteases and stability of the cofactors can determine the selection of buffers, salts, pH and additives. The ionic strength of the proteins can be used to select the precipitation techniques. The ionic strength defines the hydrophobic interaction chromatography. The need for detergent in the purification process can demand some crucial chromatographic steps. These steps will help us in the removal of the detergents. It is important to select appropriate detergent. The solubility of the proteins in the organic solvents is important to select the parameters which are needed to perform the reverse phase chromatography. The stability of the proteins with the pH value can be used to select the buffers needed for extraction. This property is also important in the purification process. We can use this property to select the parameters needed in the reversed-phase chromatography and ion-exchange chromatography. The stability of the proteins with varying temperatures can determine the selection of temperatures.

The laboratory in which the protein is purified should be effective and need to be assessed properly. The laboratory must have access to assays that can determine the purity of the proteins. The lab should have reliable assays that can yield desired proteins. In the case of protein purification on large scales, assays needed to remove the impurities must be considered important. The sample preparation before the chromatography procedure is based on the type of sample.

1.3 CHROMATOGRAPHIC TECHNIQUES

If we need to isolate a protein from a crude mixture, it is essential to know the physical and chemical nature of the protein. Though, there is no simple way to isolate proteins of all kinds.

The processes involved in the purification of the first protein may denature or deactivate the second protein. It is important to set the goal prior to the isolation process. The purity range can be based on the purpose to which the purified protein serves. Chromatography includes a class of separation techniques that includes molecular retardation along with the progressing solvent.

This technique is being used in several protein separation techniques. Among the chromatographic techniques, column chromatography related physical design is common to this day. In this technique, the stationary phase will be packed in a tube.

This is called a column. The eluent being used is called a mobile phase and it is mostly pumped. The rate at which the molecules get adsorbed on the stationary phase will decide how the mobile phase will behave. The chromatography assisted protein separation has evolved as an effective tool in the protein purification process.

The properties of a protein must be considered while isolating them from other contaminants. We can observe small variations in their biospecific interaction, hydrophobicity, charge and size. The first step in the chromatography technique is capturing. The product will be attached to the adsorbent. The impurities in the solution will be allowed to pass as they are not attracted to the adsorbent. Most of the proteins that are weakly bound will be washed away. The experimental conditions will be modified so that the desired protein can be extracted by the elution buffer. There are many types of liquid chromatography. The stationary phase in them will be varying. In the case of size exclusion chromatography, we need to consider the sieving properties exhibited by the stationary phase. In this technique, we do not consider the adsorption process. The column chromatography can be classified into

- Ion Exchange Chromatography (IEXC) (Figure 1.7)
- Size Exclusion Chromatography (Figure 1.8)

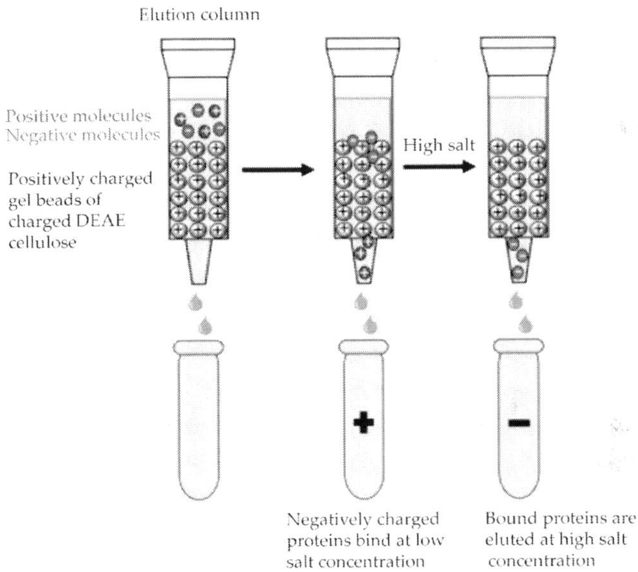

Figure 1.7. Ion Exchange Chromatography.

Source: https://www.sciencedirect.com/topics/immunology-and-microbiology/ion-exchange-chromatography

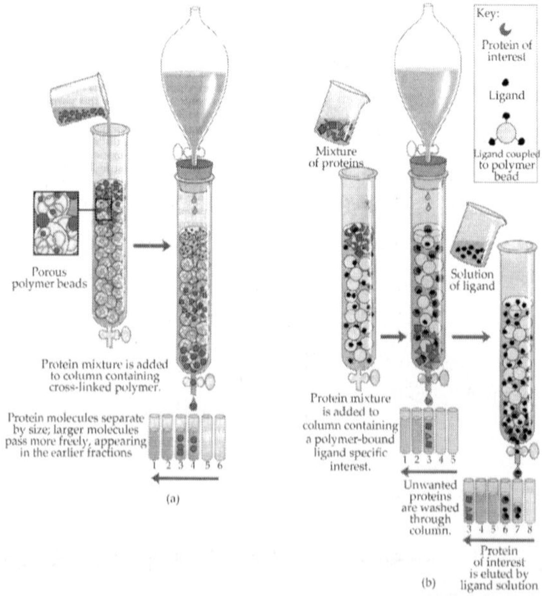

Figure 1.8. Size Exclusion Chromatography.

Source: http://www.bioinfo.org.cn/book/biochemistry/chapt06/bio1.htm

The ionic interactions can be important in the case of ion-exchange chromatography. The separation of proteins will be based on the surface charges. The proteins can be classified as complex ampholytes. They can have positive and negative charges. The isoelectric point will be based on the amino acid residues present in the structure. Mostly, the positive charges can be due to the presence of histidines, lysines and arginines. This can be based on the buffer pH. If the pH value is less than 8, then the free N terminal amine can give the proteins a positive charge. The negative charges on the protein are due to glutamate and aspartate residues. The carboxyl group at the C terminal will impose a negative charge on the proteins. Most of the residues will be prone to ionization if the pH is greater than 6. In the case of high pH, there will be ionization of cysteines. The groups which are charged will exist on the surface of the protein. It is convenient to separate proteins with the help of a cation exchanger, fixed negative charges, anion exchanger and stationary phase. The cation exchange chromatography can be performed when the pH value is less than the isoelectric point. In the case of anion exchange chromatography, the pH value will be more than the isoelectric point. (Figure 1.9).

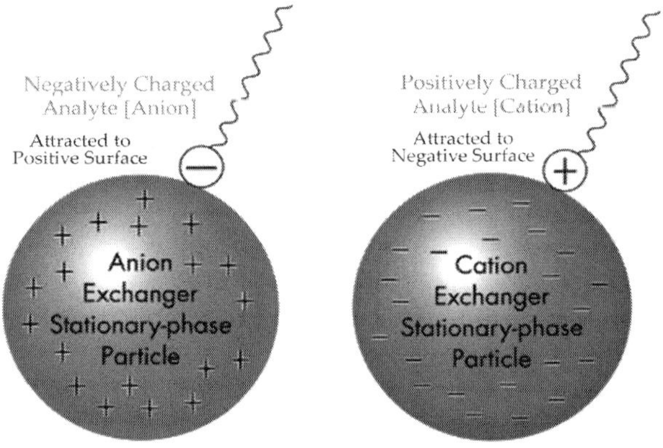

Figure 1.9. Cation & Anion exchange chromatography.

Source: http://technologyinscience.blogspot.com/2011/09/ion-exchange-chromatography-principle.html

The ion exchange chromatography will be based on a certain pH range which helps the protein to be stable. In order to achieve good adsorption rates, the pH of the buffer solution being used must be slightly greater or lower than the isoelectric point so that the desired analytes will be separated. The pH of the ion exchanger may not be the same as the buffer pH. This can be called as 'Donnan effect'. (Figure 1.10)

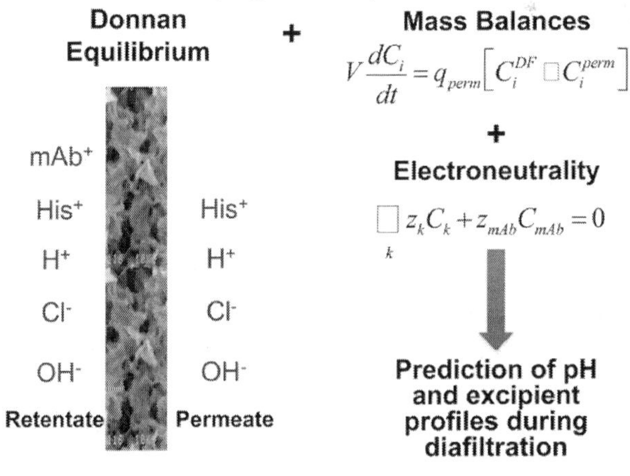

Figure 1.10. Donnan Effect.

Source: https://onlinelibrary.wiley.com/doi/abs/10.1002/biot.201800517

The protons can be expelled or in some cases attracted from the matrix. This characteristic of the protons will be based on the charge residing on the matrix.

The pH in the vicinity of the matrix will be one unit lower in the case of cation exchangers and it is one unit higher in case of anion exchanger. For example, if adsorbed protein on the cation exchanger is at pH 5, then we can assume that the protein will be exposed to pH 4. At this pH, the proteins are prone to deactivation and denaturation.

Most of the proteins will exhibit a negative charge if exposed to a pH range from 6 to 8. It is important to note that the anion exchangers must be used in the case of ion-exchange chromatography.

We need to maintain a large variation in charge between the contaminants and target protein. We can use IEXC several times in the purification process as the pH can be modified.

The relationship between the ion exchanger and protein can be based on the ionic strength and net charge. It is also based on the charge distribution on the protein surface. In case there is a high concentration of groups that are charged, then there will be protein binding.

This occurs even when the charge is zero. The chromatography technique can be based on the confirmation of the protein and so any structural change can affect the separation of proteins.

The ion exchanger can affect the separation process. Most of the ion exchangers are categorized into weak and strong. This can be based on their pKa values.

The change in the pH values will not change the charge existing on the ion exchanger. Most of the weak ion exchangers can support functional ionic groups.

Some of them are diethyl ammonium and carboxylate. In the mobile phase, the conductivity and pH of the buffer are important. These factors affect the adsorption process. The conductivity should be low and pH must be selected so that there will be an optimal charge on the protein whilst the separation process.

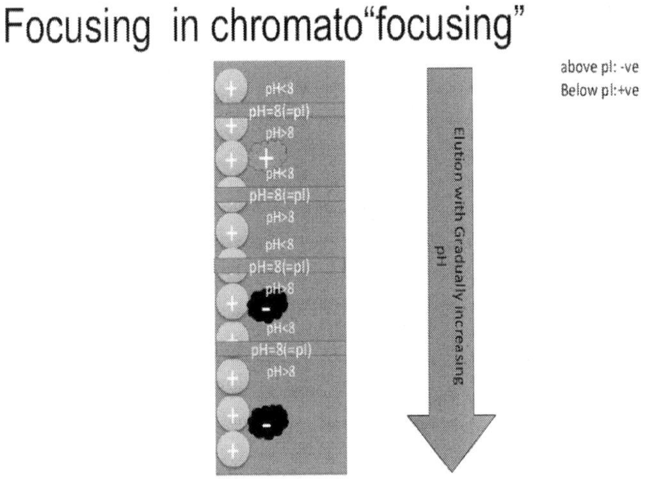

Figure 1.11. Chromatofocusing.

Source: https://www.slideshare.net/amlano/chromatofocusing

The selection of the buffer ions can also affect the protein separation process. In case, there is the same charge on the resin and protein, then the protein will pass via the column and there will be no noticeable adsorption. The reduction in the affinity can be based on the buffer changes. We can increase the ionic strength with the addition of sodium chloride. The lowering of buffer pH can be used in this process. With the aid of improved systems which have a high buffering capacity, we can get proteins with high resolution. This can be done with the pH elution technique. This can be called as chromate-focusing (Figure 1.11). The ions produced will compete for the binding sites that are present on the resins. The proteins which are having weak charge can be eluted with dilute solutions and the proteins which are strongly charged requires concentrated salt solutions. The properties of the buffer can be changed in the case of gradient elution.

1.4 ELECTROPHORETIC TECHNIQUES

The electrophoresis process was devised 60 years ago by Dr, Jack Williams. The proteins are classified as functional units in the cell and they give us information about health and disease. Most of the proteomic techniques were outcasted by the electrophoretic techniques. In the process of protein electrophoresis, there will be the migration of the charged molecules under the electric field. The mobility of the proteins will be based on various factors. They are charge supplied on the surface, strength, shape and size of the electric field. The pH of the buffer may affect the migration of the proteins.

The net charge of the proteins will be changed. The separation of the samples will be done with gel support. The gel is placed or immersed in the buffer. The gel will connect the two electrodes. The proteins loaded in the electrophoretic wells can migrate to the electrodes. If they are anions, then they can migrate toward the cathode. Post-separation, the gel will be subjected to the staining process.

The staining process needs a variety of chemicals so that the proteins which are separated will be visualized. Later, we can detect the fluorescent tags and radioactive tags.

Then immunoblotting will be done. We can now determine the polypeptides which are separated. The moving boundary methods have become crucial in the advancement of the electrophoretic methods (Tiselius A, 1938). Some advances from the moving boundary methods are the paper and gel electrophoresis.

Tiselius separated proteins from the human plasma and then there was no supporting medium such as a gel. In the moving boundary method, there will be a production of heat. The protein bands will be diffused and these bands have only heads (Righetti PG, 2005). Due to this limitation, there were attempts made to use solid mediums.

The solid mediums were polyacrylamide, starch and paper. These solid mediums stopped the sedimentation process whilst protein separation. Paper was considered as the first solid medium attempted. Snake venom was separated using paper as a solid medium (Von Klobusitzky *et al*, 1939). Amino acids were separated using cellulose-based filter paper as support media (Evered DF, 1959).

In the case of paper electrophoresis, there wasphysical separation between the electrophoresis compartment and electrode chambers. This was done to avoid contamination. The identification of single amino acid substitution in the haemoglobin as a cause of sickle cell anaemia was done by paper electrophoresis and chromatography (Ingram VN, 1958). (Figure 1.12 and 1.13)

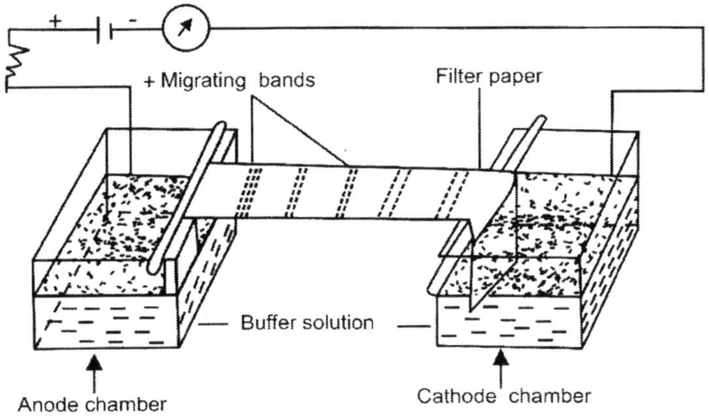

Figure 1.12. Paper Electrophoresis.

Source: https://www.slideshare.net/TapeshwarYadav1/paper-electrophoresis

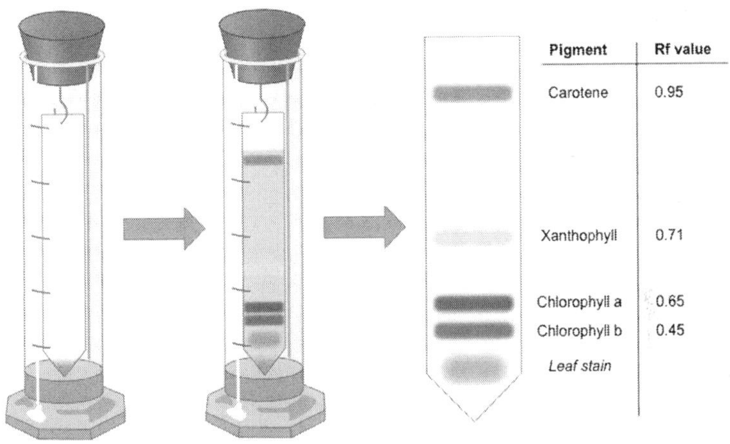

Figure 1.13. Paper Chromatography.

Source: https://microbenotes.com/paper-chromatography/

The gels made out of starch were used to separate haptoglobins and this was considered as an advancement in the gel electrophoresis (Smithies O, 1955). It is this process that introduced the sieving concept in the electrophoresis. Discontinuous buffers were used in the starch gel electrophoresis process (Poulik MD, 1957). Polyacrylamide support matrix was first used in 1959 (Raymond *et al,* 1959). The disk gel system was designed in 1964. The copolymerization of bifunctional cross-linking agents and acrylamide produced polyacrylamide gel matrix (Gelfi *et al,* 1986). The cross-linking agent used was Bis or N,N'-methylenebisacrylamide (Gelfi *et al*, 1986 and Righetti *et al,* 1981). This linking reaction was mediated by ammonium persulfate and TEMED or N,N,N,N-tetramethylethylenediamine (Andrews AT, 1985). The main use of this process is that the pore sizes can be controlled by altering the cross-linker and acrylamide ratios. The pore size can be considered as a vital factor since it determines the protein migration distance and molecular sieving capacity of the gel (Maurer Hr, 1978). The acrylamide concentration defines the pore size of the polyacrylamide gel. If the acrylamide concentration is high, then we can separate the proteins with low molecular weight. The porosity gradients to separate the protein samples into demarcated bands increased the capacity of the PAGE (Margolis *et al,* 1967). The separation of proteins has become easier and effective after the introduction of SDS or sodium dodecyl sulfate (Shapiro *et al,*1967). This chemical can denature proteins and thus unfold the protein structures. Due to this the variations in the secondary structure and tertiary protein structure will be reduced and so there will be polypeptides instead of proteins. PAGE can be operated at a wider pH range i.e. from 2 to 10 (Andrews AT, 1986). The high ionic strengths can yield lower diffusion coefficients and so there will be effective protein separation. It is important to use buffers that have high ionic strengths. The resolution will be greater if we use discontinuous gel and the relevant buffer system. This system was devised by Ornstein and Davis (Ornstein L, 1976 and Davis BJ, 1964). This system is widely used today. The effective version of this system is Laemmli SDS-PAGE system (Laemmli UK, 2011).

In the case of a discontinuous system, voltage and pH gradients will be employed in gel compartments. There will be lower resolving gel and upper stacking gel. The pH of the stacking gel will be less than the resolving gel. In the stacking gel, the size of the pores is large. Due to ionic strengths, pH, sample preparation buffer and large pore size, there will be thin and sharp starting zones (Andrews AT, 1986). We can use beta-mercaptoethanol and SDS to break the disulfide bonds. This chemical can solubilize and denature the proteins. In order to visualize the migration of the proteins, Bromophenol Blue will be used. In the stacking gel, Tris/HCl buffers are used and pH will be maintained at 2. In the electrophoretic run, Tris/glycine buffer will be used so that the protein molecules will be separated as sharp bands. These sharp bands will be between the glycine ions at the trailing end and chloride ions at the leading end. For the resolving gel, 10% acrylamide will be used to ensure a small pore size. There will be higher salt concentration and pH will be higher. When the protein bands reach the resolving gel, the glycine will be ionized. There will be no voltage gradient and so the protein separation will be based on the size. The proteins can be separated by SDS- PAGE and can be visualized with metal depositional methods or chemical staining methods. There are some steps that are to be followed whilst staining procedure. They are 1. Elimination of the substances which resists staining, 2. Staining with reagents, 3. Destaining to halt the staining process (Steinberg TH, 2009). Coomassie dyes can be used to stain the amino acids (Neuhoff *et al,* 1985). Silver staining methods can be effective thandyes (Harris et al., 1965).

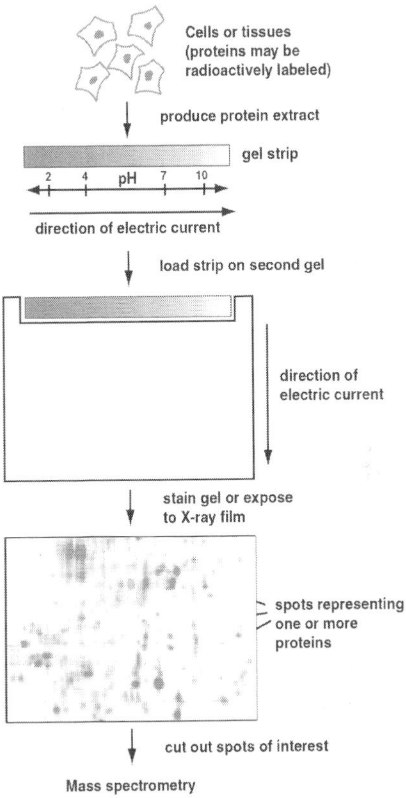

Figure 1.14. Two-Dimensional Gel Electrophoresis.

Source: https://www.researchgate.net/publication/236181452_Proteomic_Approaches_for_Studying_Alcoholism_and_Alcohol-Induced_Organ_Damage/figures?lo=1

2-DE or two-dimensional gel electrophoresis (Figure 1.14) was considered as an important electrophoretic technology before mass spectrometry. In this process, the protein separation will be in two dimensions. The first dimension is based on the isoelectric points and the second is based on the molecular weights. The two-dimensional gel electrophoresis can be used to separate proteins from the blood and this technique can be used in evaluating the tissue proteins and cells. The stable pH gradients can enhance the resolving power and so we can easily identify the proteins that are in lesser concentrations. There are several techniques that can be used to visualize the proteins, which are separated. We can use silver staining, fluorescent labeling and radioactive labeling.

1.5 CONCLUSIONS

The selection of the purification process can be a tedious process. The protein purification process involves both extraction and purification steps. There are many procedures in the extraction process that mainly include the disruption of cells. The animal tissues and cells may be disrupted smoothly. We can use vigorous force on the plant tissues to cleave the cell envelopes. There are many procedures that take part in the purification process. They are chromatography, density, size and solubility. There are various chromatographic techniques which are being used in the separation process. Some proteins can be isolated and purified with simple procedures. The electrophoresis technique is widely used in the purification process. There are many protein detection techniques that are in use now. There are many advantages and disadvantages connected with a single method. These methods can be used for both qualitative and quantitative analysis of proteins. These techniques are based on specific structures present in proteins calledamino acids. These amino acids will have the free amino group. It is important to follow a protein-specific manner. If we want to characterize albumin with one electrophoretic gel, then we need to compare it with the same protein on another gel.

Protein analysis has helped us to understand the biomedical sciences and biological sciences with high resolution. The gel electrophoresis and chromatographic techniques have helped us to know the internal constituents of the cells. In the case of electrophoresis, the liquid phase is driven by the electric field and in chromatography, the pressure is used. The analytical protein electrophoresis can be used to understand the biochemical characteristics like protein mass and isoelectric points. These techniques can be used to quantify a particular protein from a mixture. Immunoprecipitation can be considered as an important technique in the purification of proteins. The preparative electrophoresis can be used to separate proteins from a mixture and we can even clean samples so that the trace amounts of contaminants can be eliminated. 2DE or two-dimensional electrophoresis, SDS-PAGE and isoelectric focusing can be considered as important electrophoretic techniques. These analytical techniques can be scaled up and may be upgraded to the preparative level.

2

Equipment and Strategies for Protein Purification

CONTENTS

2.1 INTRODUCTION

The advancements in the genomics field have provided us different platforms for observing the molecular interactions occurring in the inter environment of the cells. The genomes have the information needed for protein production. The genome cannot define the function of the protein post-production. The proteins produced can function at different sites and have diverse actions. So, knowing the genome does not guarantee complete protein information. There are many post-translational modifications that occur inside the cells with the aid of cofactors. These modifications determine the activity of the proteins. It is important to understand the complete proteome and the techniques connected to the study of the proteomes. Proteomics helps us to know important proteins and assists us in their isolation.

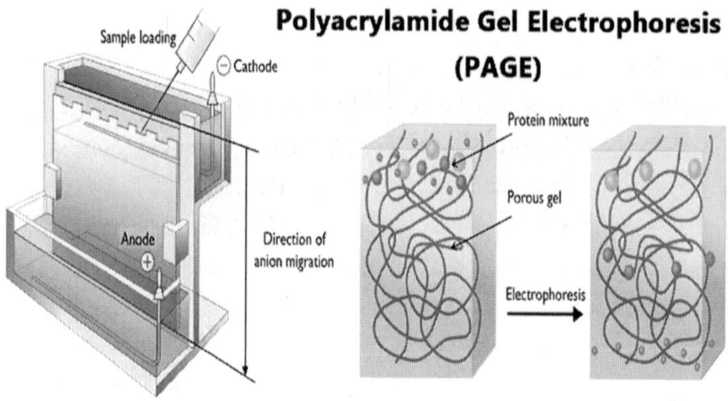

Figure 2.1. 2-Dimensional Polyacrylamide Gel Electrophoresis.

Source: https://microbenotes.com/polyacrylamide-gel-electrophoresis-page/

2-dimensional polyacrylamide gel electrophoresis of 2D PAGE (Figure 2.1) classified as an important technique in the identification of proteins. In this technique, proteins do separate on basis of their charges. Then the separation will be based on the molecular mass. We can identify polypeptides using mass spectroscopy. It is important to know the information available about the proteins through archive data or other sources. We should check for purification protocols that are related to the desired protein. Knowing the protocols in advance can save more time. Searching for purification techniques give us technical information so that the entire process will be systematic. It is better to use a computer-based purification program prior to actual lab work. GE health care simulation and Proteinlab of Leeds

University helps us in this process. These programs will help us to anticipate the problems well before the use of chemicals. With these programs, we can know the empirical nature of the purification process with fewer expenses.

It is important to know the fundamental biochemical parameters. We need to establish Michaelis-Menten constant or K_M. We need to know the effects of inhibitors and activators on the function of the desired protein. The post-translational and molecular mass can be defined by the protein after purification. The sequence of the protein can be used to predict the corresponding gene sequence. We can grow crystals of the purified proteins. We can produce antibodies from the purified proteins. We may know the cellular location of different species. We cannot produce a protein that is completely pure as it will be expensive. It is important to decide the purity level well before the start of the purification process. In therapeutic applications, we may need a homogeneous product. We need to minimize the steps involved in the purification process without compromising the yields. Each stage will reduce the protein activity. Apart from the number of steps involved in the purification process, there are other factors that may reduce the protein activity. The proteolytic activity can be considered important whilst understanding the loss of protein activity. In the case of cells from the eukaryotic samples, there will be compartmentalization of organelles. When we extract these organelles, the proteins present in different organelles will get mixed.

There will be proteolytic enzymes that are present in the lysosomes in the case of animal cells which initiate proteolytic activity. In the case of plant cells, the lysosomes are present in the vacuoles. We need to know the proteolytic enzymes present in the samples so that we can add inhibitors. In the initial purification steps, the protein concentration will be higher. There are several proteins that are prone to the proteolytic enzymes and in some cases, they may not affect the desired protein. If the percentage of protein in the total protein component increases, there will be a breakdown of the target protein. We can reduce bacterial contamination by using filtered chromatographic buffers. The buffers must be filtered via 0.2 micrometer filter. We should centrifuge the extracts regularly. The centrifugation is done for 20 minutes at 13000 g. The proteolytic assay used to check the co-purification. In case, we get positive proteolytic assay, we can go for certain modifications to reduce co-elution and we can even add inhibitors for that matter. The optimum temperature range to reduce the proteolytic activity will be 5 to 10°C. The stability of the protein varies with changing temperatures and pH. Once we know the proteolytic profile, it is important to establish the stability conditions for a purified protein. The extract will be subjected to several temperature ranges and pH values. With this step, we will know the pH and buffer compatibility. If the concentration of the protein is low, then the proteins bind to materials like resins, plastic and glass. This can be avoided by Triton X-100. This process is called as silanizing. The desired protein can be reduced at air-liquid interfaces. This reduction is mainly due to the insoluble and soluble aggregates and it is mandatory to lessen the factors that initiate the foaming process. The divalent metal ions can be considered as important entities in the protein purification process. In the proteins extracted from the cells, there will be several covalent disulfide bridges. These bridges protects the tertiary structure. The divalent metal ions can attach with the sulfhydryl groups which were reduced. This will leave the enzyme inactive. We can use EDTA to eliminate this problem as EDTA can bind to the divalent metal ions.

2.2 PROTEIN STRUCTURE

There are 20 amino acids that makes proteins with diverse functional groups. The protein sequence will start with an amino acid which has amino terminus. The sequences will end with the carboxyl terminus. This linear structure is called as primary structure of a protein. The primary structure undergoes a folding process to adopt a secondary protein structure. In the secondary structure of a protein, the energy states will be lowered. Alpha helix and beta strands are the major secondary structures in the folding process. Four levels of protein structure are given in figure 2.2.

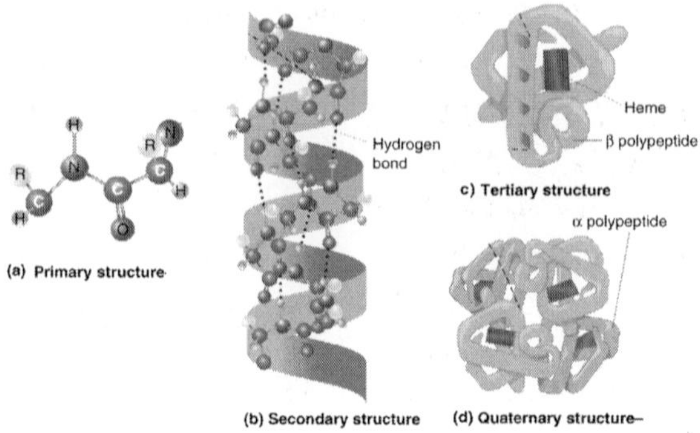

Figure 2.2. Levels of Protein structure.

Source: https://www.mun.ca/biology/scarr/iGen3_06-04.html

The ribosomes will harbor the protein production process. The exact conformation of the proteins will be maintained by chaperones or helper proteins. These helper proteins can be used to position the amino acids with functional groups (hydrophobic). The tertiary structure is stabilized by noncovalent forces. Some of the noncovalent forces are salt bridges, Van der Waals forces, hydrogen bonding and hydrophobic interactions. The hydrophilic portions of the amino acids are in the water. There are some proteins that are present at the vital point of several metabolic pathways. These proteins exist as superstructures. The proteins in this form can be called as quaternary structures. The charge on the protein surface are attributed to the amino acid side chains. These amino acids behave like weak acids. This is based on pH values. Histidine or H are positively charged if the pH value if less than 5.5. The amino acids such as arginine or R and lysine or K holds some charge above pH 9. In the case of oppositely charged amino acids, there will be the formation of salt bridges and so the tertiary structure will be stabilized. The charge present on the protein surface can be based on pH. If the pH values are high, then there will be a negative charge. If the pH is low, then we can expect a positive charge. At certain pH, the charges will be balanced and so the charge will be nullified. This pH value is called as an isoelectric point. We can designate isoelectric point with 'pI'. As the proteins evolve from different genes, there will be different numbers of amino acids. There will be different side chains with charges. This will result in diverse pI values. The charge difference can be used in protein purification techniques.

Eight amino acids have different hydrophobic properties and it is based on the functional groups. The variations in the surface hydrophobicity can be used to extract the desired protein from a mixture. The important amino acids are methionine or M, alanine or A, valine or V, isoleucine or L, phenylalanine or F, tryptophan or W and tyrosine or Y. These amino acids are present in the pockets on the protein surface. Most of the proteins are soluble in water. The hydrophilic nature of the proteins is based on the amino acids. The solubility of the proteins in water can be based on ionic strength, pH and temperature. When the salt concentration increases in the solution, the proteins exhibit differential solubility. This is called as 'salting out'. If the salt concentration is low, then the proteins exhibit increased solubility. This is called as 'salting in'. These processes can be useful in fractionating protein mixtures. Most of the proteins exhibit biological functions. They are stereospecific. The biospecificity exhibited by the proteins can be used in the protein purification process. This property can be used to isolate protein. Agarose linked with protein G can be exploited to purify the IgG antibodies from the serum samples. There are several proteins that have different amino acid compositions and they have varying molecular mass. We can use the size difference to isolate protein from the complex mixture.

Post translation, there will be some additional groups that can change the amino acid conformation. One-quarter of the proteins present in the cell can have tyrosine, threonine and serine residues linked with phosphate groups. The sugar groups will be attached to serine and arginine. Most of the modifications change the surface charge of the protein. This feature can be used to isolate the protein from a mixture. It is easy to produce the desired protein from different organisms. There are several techniques used in the purification process. Some techniques will use the protein surface charge. Some of the important techniques are Isoelectric focusing or IEF-Preparative, polyacrylamide gel electrophoresis, hydroxyapatite, chromate focusing and ion-exchange chromatography. Some techniques use the hydrophobic character. They are reverse phase chromatography and hydrophobic interaction chromatography or HIC. The biospecificity property can be used by various techniques. They are Immobilized metal affinity chromatography or IMAC, immunoaffinity chromatography, covalent chromatography and affinity chromatography. The techniques that use the molecular mass feature are SDS-PAGE and size exclusion chromatography or SEC.

There are about 30000 genes present in the human genome and these genes will encode about 400000 proteins. These proteins are derived from about 20 amino acids. The purification strategy must include the combination of techniques that exploit size, hydrophobicity, charge and biospecificity. The purification procedures must prevent overlapping techniques. In the case of one step purification, we can exploit the biospecificity alone. The affinity ligands can be costly and they need rough elution conditions. Affinity resin can be used after an enrichment step via HIC or IEX. The resins must be studies with pilot-scale experiments. The post-translational modifications such as phosphorylation and glycosylation can be used in the production of the proteins. There are 4 amino acids that can exhibit charges at physiological pH. The conditions that favor the binding process and the total concentration of the desired protein must be determined using small scale experiments. We can observe the oxygen -free environment in the cytoplasm. The concentration of protein in the cytoplasm will be higher. There will be a dilution of the protein. The diluted protein will be subjected to the oxidizing environment. The first step in the protein purification process is to remove water and inhibitors. The high-resolution protein purification protocols include techniques such as PAGE, IEF, chromatofocusing, immunoaffinity chromatography and affinity chromatography. At the final stage of the protein purification, the concentration of the desired protein tends to drop.

2.3 CHROMATOGRAPHY- THEORY

Chromatography is nothing but 'coloured writing'. Mikhail Twsett used this technique to demarcate plant pigments with calcium carbonate (Eltre et al., 1993) (Figure 2.3). If the compounds are placed between two immiscible phases, they tend to distribute between 2 phases. The chromatography techniques are based on the distribution and partition between two immiscible phases. The two phases are the mobile phase and the stationary phase. Gas or liquid can be used as a mobile phase and liquid or solid can be used as a stationary phase. In the case of partition chromatography, the mobile phase can initiate new distributions between both phases. When the samples pass through the column, they continuously move between the two phases. Some compounds can get distributed via mobile phase and others will get distributed in the stationary phase.

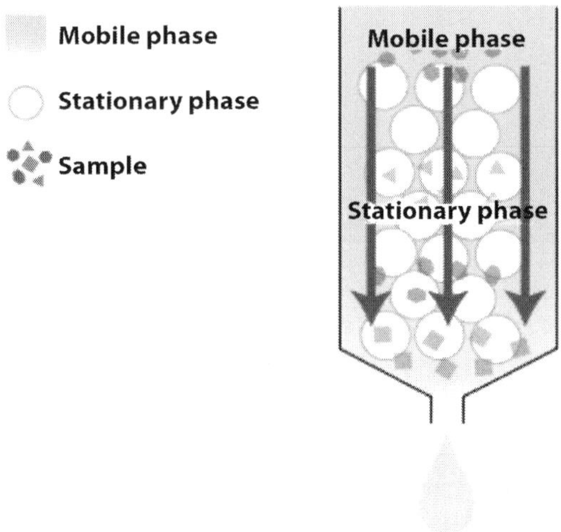

Figure 2.3. Chromatography with phases.

Source: https://www.mblbio.com/bio/g/support/method/chromatography.html

The equipment needed in the column chromatography must not be expensive and complicated. It must be kept in mind that if the purification percentage is not satisfactory, there is no need of spending more money and time on the equipment. The type of column to be used will be based on the sample concentration which is to be applied. The adsorptive techniques such as HIC and IEX need fat columns but shorter ones. In the case of size exclusion chromatography, it is preferred to use long and thin columns. Low -cost columns can be made using plastic syringes or pipettes. Flexible tubing is needed to attach the columns with the chromatography set-up. In the case of low-pressure systems, we can use polyethylene tubing. If there are large particles in the stationary phase, then we can move the mobile phase via the stationary phase. This can be done by applying hydrostatic pressure. We can adjust the flow rate by altering buffer height. We can use a piston pump or peristaltic pump to produce hydrostatic pressure. The fraction collectors must be used in the chromatographic runs. The complex protein mixture can be added to the resin which is present in the column. The volume of the samples can be divided into fractions. Most of the proteins can exhibit an absorption peak at 280nm. The main reason behind this is that the amino acids tryptophan and tyrosine are present in the primary structure of the proteins. The fractions can be analyzed at 280nm. This absorbance is designated as A_{280}. We can continuously record the fractions with the UV light. Gradient markers are important in the chromatographic processes. The chromatographic techniques such as HIC and IEX need variations in the salt concentrations. These variations can help us to elute the bounded proteins. We can add the buffer with a certain concentration via column and then the protein will be eluted. After this step, the salt concentration will be changed. We can get the fractions due to isocratic elution.

We can use the gradient marker to produce continuous variations in the salt concentrations. Almost all the chromatographic techniques follow certain steps that are interrelated. It is important to know all the information about the desired protein. Complete knowledge of the protein that is to be purified can help us deal with the problems which may appear at later stages of the chromatographic run. The initial conditions needed for a successful chromatographic run will be provided by the knowledge we gained about the protein. Most of the resins used in the chromatographic process will be stored at 4°C. We can store them in 20% ethanol. The resin can be added to the glass mesh. This can be done by applying a vacuum. There should be equilibration of the resins so that the desired protein will attach to the resins. This can be done with the help of the Buchner flask. The column used in the chromatographic technique should be inspected for any leaks. Once the bed volume is packed, the liquid flow should be halted. The column connector must be attached to the upper end. There should be no air inside the system. The column should always be packed at the operation temperature of the experiment. The temperature to be used in the process will be decided by the thermal stability of the enzyme. It is based on the proteolytic enzymes in the crude extract. The resins will have rigid structures and this will encourage good flow rates. In case the target protein is less stable, then the protein-containing solution should be maintained at low temperature throughout experiment. The samples can be applied with the aid of the sample loop. If the beads are smaller, there will be an increase in the surface volumes. We need to apply high pressures on the small beads so that the liquid can move via column. We need high -pressure systems. The flow rates are mostly expressed in terms of ml/min. The sample can interact with the resins and the proteins can associate or bind with the resins. If there are other proteins, there will easily drain via the column. These contaminants must be eluted from the column. This can be done with the start buffer. The resin in the column can be washed and the instruction manual must be followed at all times. The resins can be stored in a preservative. We can store it by passing 2 column volumes of water first followed by preservatives. The eluted fractions must be assayed for the desired protein. The peak fractions can be identified and later pooled. The protein fractions can be measured. The fold purity and percentage yield can be measured. We can use SDS-PAGE for this purpose.

A protein purification balance sheet must be used so that we can record the efficiency of the electrophoretic or chromatographic technique. This balance sheet can have information on the biological activity, protein concentration and volume of the protein present. This balance sheet can give us data on the purification degree and amount of protein present in the extract. The protein and enzymatic activity present in the crude extract explains to us the highest activity possible in the experiment. In the crude extract, the fold purity can be assumed as 1 and the percentage yield can be assumed as 100%. Acids and bases regulate the chromatographic process, so it is essential to know the facts about them prior to the chromatographic run. Bronsted has explained acid-base reactions. These reactions occur with the transfer of protons or H^+ ions. In general, Bronsted acid can be classified as a proton donor and the Bronsted base can be called as a proton acceptor. The acids can be classified as strong acids and weak acids. The classification is based on the dissociation constant called K_a. In the case of a strong acid, the proton will be released at every pH value and they exhibit high K_a values. Weak acids will exhibit lower K_a values. The K_a values are based on the pH values in the case of weak acids. All the proteins have optimal pH values at which they are active at least during their storage. When the pH values change beyond the optimal values, we can expect protein denaturation and inactivation. It is essential to maintain pH according to the protein characteristics. Any buffer solution can be prepared with a weak acid and corresponding salts which we call as a conjugate base. Buffer can also be produced with a weak base and its corresponding salts. We can use the Henderson-Hasselbalch equation while dealing with buffers.

The rate at which the chemical reaction proceeds can be doubled for every 10°C rise. The forces which can help the protein against denaturation and inactivation are hydrophobic interactions, van der Waals interactions, ionic bridging and hydrogen bonding. The Bicinchoninic acid or BCA can chelate copper ions so that there will be the development of colour. This can be called a Biuret reaction. The protein assay to be used will be decided by the purification protocol which varies with the lab. We can add-2-mercaptoethanol or 2-ME and Dithiothreitol or DTT, glutathione-reduced and cysteine to prevent oxidation. The divalent heavy metal ions can be linked with the surface of the protein or active site and this will cause problems. It is a better option to use a chelating agent like EDTA to lower the metal ion levels. Caution must be ensured while adding EDTA as it can chelate with magnesium and calcium ions. There are several inhibitors present in the eukaryotic cells. The proteinases can break the peptide bonds of the inner side so we call them endoproteases. These enzymes are classified into serine and sulfhydryl proteinases. Almost all the proteinases will be inhibited by alpha2-macroglobulin. The serine proteinases can be inhibited by soybean trypsin inhibitor, DFP or Di-isopropylfluorophosphate, PMSF or Phenylmethylsulfonyl fluoride and antipain. The sulfhydryl proteinases can be inhibited by Iodoacetimide, Leupeptin, E64 and antipain. We need to carefully choose the inhibitors so that the desired proteins may not be damaged. The nucleoproteins and nucleic acids must be removed. We can add DNase to extract the nucleic acid linked with the proteins. In the case of the ultrasonication process, the chromosomes will be sheared. The lipoproteins must be removed from the solution. Calcium chloride and dextran sulfate can be used to extract the lipoproteins. The plants release secondary metabolites such as tannins and phenols. Whilst precipitation, these secondary metabolites will be deposited. The enzymes like phenol oxidases can transform phenols to quinones. These chemicals can react with the proteins and so there will be problems in their aggregation and precipitation. This activity can be checked by enhancing DTT concentrations or 2-ME concentrations. These chemicals must be added to the extraction buffer. We need to add certain chemicals to extract the membrane proteins.

We need to add, 0.25M sucrose so that the membrane integrity will be maintained. We can add sucrose to the extraction buffer and resuspension buffers. The membrane proteins can be classified into integral proteins and peripheral proteins. The peripheral proteins will be linked to the membrane via ionic interactions and the integral proteins will be linked to the membrane via hydrophobic interactions. The peripheral proteins can be removed by high salt concentrations in buffer solution. These proteins can be purified into soluble proteins.

The integral proteins can be solubilized by solvent and detergent. The solvents must be avoided as they may affect the protein structure. These solvents can be classified as a fire hazard. We need to select an appropriate cell/tissue disruption strategy. If the correct strategy is selected, there will be a high protein yield. The cells from the fungal, bacterial, plant and animal sources can have different properties. Most of the animal cells present in the culture can be enclosed by the plasma membrane. We can use sonicators or homogenizers. The integrity of the cells can be disrupted by destabilizing chemicals such as solvents or detergents. We can use osmotic shock to disrupt cell integrity. The animal cells from the lungs and liver can be trimmed with scissors and homogenized. We can use a waring blender to homogenize the animal tissues. The plant cells can be about 100 micrometers in diameter. Most of the plant cells can be enclosed by the cell wall made up of carbohydrates. The plant tissues can be ground with sand mixed with acid. Most of the bacterial cells range between 0.7 to 4 micrometers in diameter. Most of the bacterial species can be classified into gram -positive and gram-negative bacteria. This classification is based on the gram staining technique. The gram-positive bacteria will have peptidoglycan enclosing the plasma membrane. The gram-negative bacteria will have a periplasmic space. This space can separate the peptidoglycan coat. The fungal cells are enclosed by the cell wall. It is made up of 90% polysaccharides. Chitin and cellulose are present in the cell wall. The filamentous fungi can be disrupted by agitation, sonication and extrusion. Osmotic shock method can be used to extract from the animal cell culture. Grinding with liquid nitrogen can be helpful to extract from filamentous fungi, yeast, bacteria, plant cell culture, plant tissue, animal cell culture and animal tissue. The acid -washed sand associated grinding method can be used to extract from bacteria, plant cell culture, plant tissue, animal cell culture and animal tissues. We can use lyticase enzymes to extract from yeast cells.

The plant and animal tissues must be cut into minute structures. Then we can place these structures in the mortar that is cooled. We need to add liquid nitrogen very slowly so that we can prevent quick freeze. The tissue must be ground with the pestle. In order to solubilize the proteins, we need to add a chilled buffer used in the extraction.

These tissues must be split into minute pieces even in case of acid -washed sand method. In the case of osmotic shock, the animal cells can be disrupted in the hypotonic buffer solution. The cells will absorb water and disrupt. The cellular contents will be released into the buffer solution. The tissue grinders and homogenizers can be used for different volumes. We can use 50 ml of extract. We can use the teflon pestle as it is better than the douce homogenizers. While homogenization, the tissues can be macerated into minute pieces and they are placed in the mortar. We need to add an extraction buffer which is cooled. The pestle can be useful to enhance the shear forces needed to break the tissue. The mortar must be cooled frequently as the heat produced in this process can damage the proteins. It is an important task to use ultrasonicators (Figure 2.4).

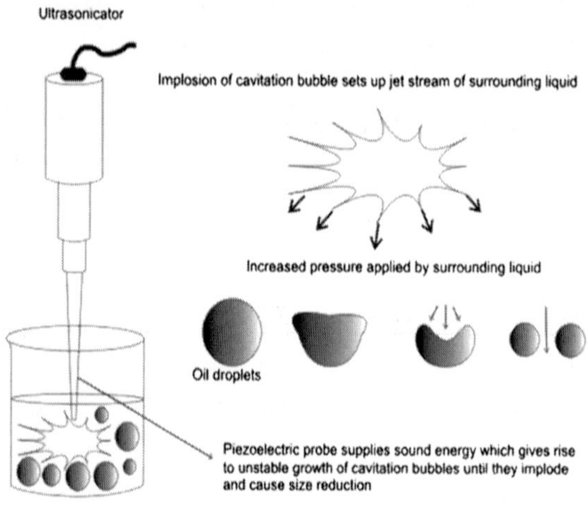

Figure 2.4. Ultrasonicators.

Source:https://www.sciencedirect.com/topics/engineering/ultrasonication

They can produce gas bubbles. Waring blenders or domestic blenders can be used to macerate the plant and animal tissues. The tissue which is macerated must be placed in the extraction buffer. We can operate these blenders for half a minute so that the tissues can get macerated. We can use polytron homogenizers to disperse the cell contents. Like the other methods, the fragmented tissues will be placed in the chilled buffer. Abrasive materials can be important in disrupting the yeast and bacterial cells. Minute quantities of yeast and bacterial cells can be fragmented by freezing with liquid nitrogen. The frozen cells can be ground with acid -washed sand. We can use mortar and pestle for this purpose.

2.4 ELECTROPHORESIS- THEORY

Post -genomic era, we find certain interesting technologies that are robust and popular. Recombinant technology has become a popular technique among others due to its success rates at least for now. Polyacrylamide gel electrophoresis is being used to identify gene products and even we can characterize them. Electrophoresis is considered as a sensitive tool to understand the protein properties.

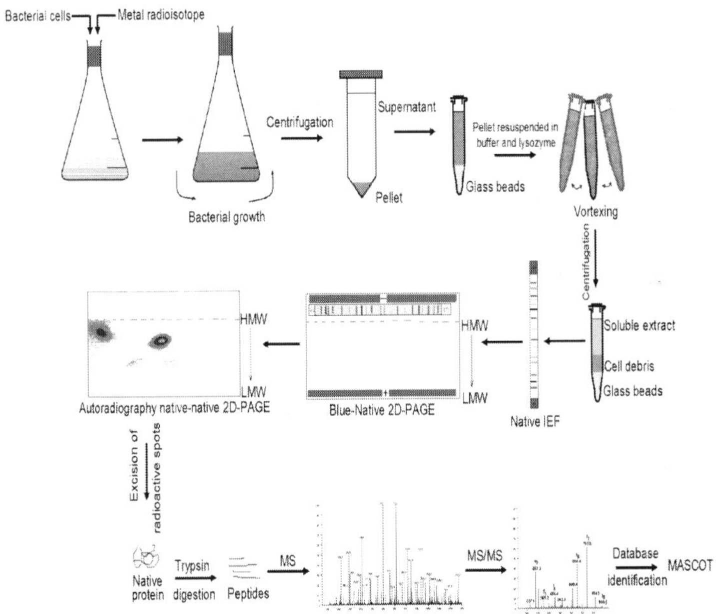

Figure 2.5. SDS-PAGE.

Source: https://www.creative-proteomics.com/services/2d-blue-native-sds-page-for-complex-analysis.htm

The electrophoresis technique is based on the principle of molecular migration or transfers according to their charges in between the electrodes using a matrix. The matrix used can be polyacrylamide gel or agarose. We can use agarose gel to disperse macromolecules like nucleic acids. PAGE can be used to analyze the purity, isoelectric point, amount and size of the proteins and polypeptides. Electrophoresis has become a promising tool in molecular biology, biochemistry and chemistry. There are several PAGE methods that are devised to solve issues with the protein purification process. No single electrophoresis method is universal as each experiment requires special parameters and conditions to be followed. 1D PAGE or one-dimensional PAGE can be classified as a simple technique and it is cost-effective. SDS-PAGE (Figure 2.5) is commonly used in determining the size of the proteins. In this process, the proteins will get denatured and if our target is active protein then this may not be the right choice. It is important to use a system that doesn't denature proteins if our target is a protein with biological activity.

To produce polyacrylamide, we need to use acrylamide and N-N'-methylene-bisacrylamide. This is also called as bisacrylamide. These chemicals can be called as monomers. The reaction between bisacrylamide and acrylamide can be initiated by chemical methods or via photochemical reactions without forming free radicals. We can call this reaction as co-polymerization and the systems used in this process that are devoid of free radicals can be called as initiators. The polymerization process can be started by tetramethylethylenediamine or TEMED. We can use ammonium persulfate along with TEMED. Methylene blue or riboflavin-5'-phosphate is used to initiate the photochemical polymerization process. The ammonium persulfate when added to water produce free radicals. These free radicals activate the monomers and, in this case, we are talking about acrylamides. TEMED can be used as a catalyst in this process. This catalyst carries electrons. The activated monomers react with other inactive monomers and hence there will be the production of long polymer chains. There will be the formation of a net of these chains as the cross-linking will be done by bisacrylamide. We can use riboflavin to produce free radicals. Ammonium persulfate can be used with riboflavin. When UV light and oxygen gas is present, riboflavin can produce free radicals via photodecomposition. The free radicals produced can be used in the polymerization process. We can use a fluorescent lamp for this purpose. Lyubimova et al (1993) described the photopolymerization systems that uses redox couple and cationic dye. We can us methylene blue as a cationic dye. The redox couple can be sodium toluene sulfinate and diphenyliodonium. The former can be used as reducer and the latter can serve as an oxidizer.

There are many factors that determine the polymerization process. There are certain factors that changes the polymerization rate and some factors determine the homogeneity, elasticity and pore size. It is important to know all these factors so that the gels can be reproduced in any lab so as to attain the same results with the same conditions. The rate of polymerization can be determined by spectrophotometry. There will be a dip in the absorbance at 260nm with the progress in the reaction. Some of the important factors that define the polymerization rate are the concentration of monomers, O_2, the intensity of light, temperature, pH, reagent purity and concentration of the initiator. Righetti et al (1981) described that the polymerization rate can be determined by the initiator type. The polymerization speed can be slow if methylene blue is used as an initiator and it will be fast if the initiator is a riboflavin system. We can expect an average polymerization rate in case of persulfate systems.

In the case of chemical -based polymerization, high initiator concentrations can yield faster polymerization rates. We need to use specific catalyst concentrations in the case of photopolymerization process. The reagent purity can be important. If there are minute contaminants in the SDS, buffers, initiators, bisacrylamide and acrylamide, then the polymerization process will be inhibited or halted. The contaminants include non-buffer ions, gel additives and metals. Ammonium persulfate solutions exhibit a high rate of hygroscopicity. It is important to prepare these solutions afresh preferably prior to the electrophoretic run. If TEMED has certain oxidation products then it appears yellow and it can be classified as less effective. Copper can halt or inhibit the polymerization process. pH can affect the polymerization rates and it is dependent on the initiators. Righetti and Cagilo showed that the persulfate system can be effective at pH 7 to 10. If pH drops to 4, we cannot expect polymerization process to continue. If pH rises to 10, then also the polymerization process will be stopped. Righetti (1993) and Gelfi (1981) described that the high temperatures can speed up the polymerization process and at low temperatures, the polymerization process will be halted. We can expect a polymerization rate at 23 to 25^{0}C. The stock solutions should be kept at 4^{0}C. The intensity of light can be considered important in the photopolymerization process. The rate of polymerization can be affected by light intensity. The rate of polymerization can be proportional to the light intensity. We need to maintain optimum light intensity. This is true for both methylene blue systems and riboflavin systems (Lyubimova and Righetti, 1993). If the light intensities are higher, the catalyst needed for polymerization will be consumed fast and so there will be no complete polymerization. In case there is weak light, we can expect polymerization for long periods. The intensity of light can be based on certain factors such as the thickness of the gel, distance from the source and light source itself. O_2 inhibit the polymerization process as most of the free radicals will be trapped by oxygen molecules. It is important to remove oxygen under vacuum. The removal of oxygen from cold solutions will be difficult. This is the reason why cold solutions will be brought into the room temperature. The concentration of monomers alter the polymerization rates. If the monomer concentrations are high, then there will be high polymerization rates.

The pore size can be controlled by acrylamide quantities and cross-linkage degrees. The degree of cross-linkage will be based on the concentration of bisacrylamide. The size of the pores will be decreased if the cross-linker proportion is increased. James et al (2012) described that 5% of bisacrylamide yield a minimum average pore size. If we need gels with high porosity, then it is better to use a 20% cross -linker.

The polymerization rates can determine the size of the pores. I Increase in the polymerization rate yield gel with less porosity. If the polymerization rate is low, then we can expect greater porosity. The capability of monomers to form polymer can affect the porosity of the gels. We can add ureas to polyacrylamide gel to disperse polypeptides and proteins. Due to this, there will be gels with small pore size. If polyethylene glycol is present in the gel solution, there will be an aggregation of lateral chains and then we expect the formation of a microporous gel.

There are many factors that affect the homogeneity of the gels. The polymerization rates control the uniformity of the gels. The persulfate-based polymerization can be sensitive to O_2. The gel mixture must be sealed so that the gel doesn't absorb oxygen. If the gel absorbs oxygen, then there will be gel which is non-homogenous nature. The thickness of the gel determine the homogeneities (Righetti et al, 1994). There will be gravity -induced inhomogeneities. The initiators of the polymerization process must be mixed properly. This will produce swirls in the gel. If we mix vigorously, the concentration of oxygen will be increased and this will affect the polymerization rates.

Hjerten (1962) defined acrylamide mixtures into T and C. C will denote the cross-linker percentage with respect to total monomer concentration. T is the total concentration of monomers in 100ml. The concentration of the acrylamide in the gel can be important in the protein separation process. Altering pH in the resolving gel can affect the molecular sieving properties in SDS-PAGE (Makowski et al, 1994). The detergents can be added to the gel systems. We can use cationic detergents such as Sarkosyl, Nonidet P-40 and Triton X-100. These detergents inhibit the polymerization process. We can use polyvinylpyrrolidone, Ficoll, methylcelluloses, dextrans and polyethylene glycol to produce gels with a large pore size (Righetti et al, 1992). Glycerol can alter the sieving properties of the gel. Glycerol can halt or inhibit the polymerization process and so there will be friction (Pennings et al, 1992).

The polyacrylamide gels can be used in slab gel systems and disc gel systems. Most of the labs use slab gel systems. Powerpack can be used to run the electrophoresis unit. The supply must be 500V and 100mA. The polyacrylamide gels can be prepared with a linear gradient maker. We can use a peristaltic pump for this purpose. The denser solution can be placed in the compartment that is downstream. We can use a 15W daylight fluorescent lamp. The distance from the lamp must be 10cm. This is important in the case of riboflavin linked polymerization. The zone electrophoretic systems have buffer ions and running buffer at constant pH. These systems can be called as continuous buffer systems. The samples will be loaded onto the wells of the resolving gel. The discontinuous buffer systems have different buffer ions operating at different pH conditions. Generally, the samples are loaded on the wells of the stacking gel. This gel can be placed on the resolving gel.

The gel systems can be classified into dissociating gel systems and non-dissociating gel systems. In the case of non-dissociating gel systems, the native protein properties will be preserved and activity intact. In the case of dissociating systems, the proteins will be denatured into polypeptides. The polypeptide composition of the samples will be studied with these systems. In the case of a non-dissociating gel system, the separation can be based on the protein size, charge and shape. We cannot estimate the molecular weight accurately for many proteins with this gel system. It is difficult to identify a particular protein only by determining the molecular weight. It is important to use additional identification techniques to determine the proteins (Tyagi et al, 1993). The buffer pH will be based on the isoelectric points of respective proteins to be used in the electrophoretic runs. The SDS-PAGE can be classified as a discontinuous gel system. It can be used in protein analysis. The protein mixture can be denatured at 100°C in the SDS-PAGE system. The heating of the proteins can be done in the presence of thiol reagent and SDS. The proteins can be denatured into polypeptides. These polypeptides can be linked with SDS. The proteins can be separated in the gels of appropriate porosity with respect to their sizes. In this system, the samples may not be loaded onto the wells of the resolving gels. In these systems, the stacking gels comprise of different buffer ions. We can use different pH. We can use SDS-PAGE in the western blotting process. The size and number of polypeptide subunits can be analyzed with this technique. We can monitor the integrity of the proteins and can compare the composition of the polypeptides. We can quantify the proteins. We can estimate the size and purity of the proteins.

There are certain proteins that can get precipitated in the stacking gels. This precipitation is due to the acidic pH of the gels. The protein samples will exhibit stacking gels. The SDS-PAGE can be run on the gradient and uniform gels. SDS yield protein precipitation and protein aggregation. We cannot expect high resolution with this technique. We use SDS-PAGE for many experimental purposes. Some of them are western blotting, to estimate the size and number of polypeptide subunits, compare between polypeptide samples, analyzing protein integrity, quantification of proteins and estimation of size and purity of proteins. There are certain proteins that get precipitated in the stacking gels. This precipitation is common at low pH. We can observe the stacking effect. The proteins should be processed by denatured it at 100°C before separating them on SDS-PAGE. The heating process must be done with both thiol reagent and SDS in excess. These chemicals can speed up the cleavage of disulfide bonds. After denaturation, the polypeptides will get linked with SDS and it is based on weight. This attachment is not based on the amino acid composition. The sequence of proteins cannot be considered in this process. We can observe rods with varying lengths. The length of the protein can represent the molecular weight of the proteins. We can observe that the polypeptides formed have a negative charge. Through this technique, we can separate the proteins according to the size.

The chemicals bisacrylamide and acrylamide can be classified as neurotoxins. They are categorized as carcinogens and utmost care must be taken to avoid unnecessary exposure. These chemicals when used in the electrophoresis process can be stored for a year. We can prepare these stock solutions by adding distilled water. It is important to note that these solutions must be stored in dark and temperature should be kept at 4°C. We need to prepare a 10% SDS stock solution. It is important to warm the solution so as to ensure that SDS is dissolved properly. The SDS solution can be stored for several days only if kept at room temperature. TEMED can get oxidized easily and it is highly hygroscopic in nature. Water will be absorbed and so there will be oxidation. It is not advisable to store TEMED for longer periods. We store TEMED for 6 months in a glass bottle kept at dark conditions. Ammonium persulfate is hygroscopic in nature. This chemical will be cleaved once it gets in contact with water. The initiator riboflavin can be stored at room temperatures. This chemical can be stored for one year. Methylene blue and sodium toluene sulfinate stock solutions can be prepared with distilled water. We can store these stock solutions in the dark for one week.

SDS-PAGE technique is considered as a popular technique that can be used to analyze polypeptides. It can be performed with an SDS-PAGE discontinuous buffer system. The resolution is higher and it can be achieved through stacking gel. The pH values of stacking gel, running buffer and resolving gel are different. The pH of the resolving gel is 8.8, running buffer is 8.3 and the stacking gel is 6.8.

The gradient gel electrophoresis is having several advantages and these are due to pore limit PAGE. The proteins can attain the pore limit in these gradient gels. It will occur until the migration rate becomes zero. At this point there will be no migration whatsoever even electrophoresis is continued and, in some cases, there might be some sort of migration (Lambin et al, 1986). The amount of sample needed is based on the concentration of proteins present in the mixture. It is also based on the detection methods and polyacrylamide gel which is to be used or we can say format. In the case of Coomassie blue staining, we can use samples of less than 2 micrometers. In the case of continuous buffer systems, the volume of the sample must be less. The depth of the well on the gel can determine the sharpness of the protein band. The stacking effect can sharpen the samples. The sample will be mixed with the buffer and this is followed by heating. If the protein samples are dilute for the electrophoretic process, there are certain solutions to this issue.

We can go for lyophilization of the samples. The lyophilization time can be based on the volume of the sample. This is also based on the salt content. It is important to know that the samples that have 0.5M should not be used and they must be contaminant free. We can use ultrafiltration for this purpose. In the process of ultrafiltration, the sample volumes get concentrated enough so that the large molecules can be formed. There will be no change in the salt concentrations. The time needed depends on the size of the membrane, concentration of the proteins and volume of the samples. The recovery of the proteins can be based on the retention capacity of the filter. We can use this method for samples that have high salt concentration, urea and other detergents.

The ammonium persulfate precipitation can be an important aspect that needs to be understood. The time needed for its precipitation depends on the volume of the samples. There are some proteins that exhibit low recoveries. Dialysis can be done opposite polyethylene glycol. This process of dialysis lower the salt concentration and sample volume. There will be some amount of protein loss. This loss in the protein concentration can be attributed to membrane retention. We can precipitate with TCA so that there will be protein concentration. In this method, we can eliminate the slats. The recovery of the proteins is quite low. We can precipitate with dyes and these dyes can be used in the protein assay (Marshall and Williams, 1995). We can use phenol-ether precipitation and so there will be more protein recovery from the solutions (Sauve et al, 1995). The proteins can be solubilized with the sample buffer. The detection of the disulfide linkages can be considered critical in knowing the structure of the subunit of the proteins. If the denaturation process takes place without a reducing agent, there will be disulfide bond formation. These intermolecular bonds may not be present in the original protein (Kumer and Davidson, 1992). It was observed that the duration and temperature in the denaturation process are not important.

If we need to know the molecular weight of the polypeptides, then it is important to add proteins with the known molecular weights. The standard value of the proteins being used must be important. In the case of a continuous buffer system, the sample must be loaded onto the starting zone that is thin. The SDS molecules can bind with the hydrophobic regions (Tanford, 1970). When the ionic strength is increased, then the SDS binding with the hydrophobic regions will be reduced. In order to determine the molecular weight, the polypeptides must bind to SDS. The unknown proteins and known proteins can be loaded onto the electrophoresis chamber. We can estimate the molecular weights by understanding the relative mobilities (Shapiro et al, 1967). The ionic strengths if higher can lower the binding capacity of SDS with the proteins (Tanford, 1955). This unequal binding of SDS molecules with the protein can interfere with the molecular weight determination (Odense et al., 1975).

If the samples don't remain in the wells, then it can be assumed that the sample density is less than the buffer. This happen due to the omission of sucrose or glycerol in the sample buffer solution. We observe a yellow color in the sample. This shows that the sample has a low pH. We need to adjust pH by adding Tris so that the sample will return to blue color. Caution must be ensured to avoid problems with Tris in the run. In some cases, we may not track the dye in the gel. It is important to ensure that there is continuous flow of current. Bubble formation on the electrodes must be prevented. A proper connection should be made between the power supply and the electrodes. In case, if we observe normal current reading, then it is important to know the polarities and check for abnormalities. If protein is not entering into the gel, then there may be certain issues that need to be addressed. The failure of protein to enter the gel can be attributed to the protein aggregation. There are some proteins that are more sensitive to the high concentrations and so there will be a increase in the rate of aggregation. This can also be attributed to the concentrating effect produced by the stacking gels. In this case, it is important to use dilute samples, especially in the continuous buffer system. Distorted bands may appear in the electrophoretic chambers. In the electrophoretic run, heat can be produced due to resistance offered by the gels. The mobility of the ions can be increased with an increase in the temperatures. There will be the formation of the temperature gradient. The gradient will extend from the center of the gel to its surface. This can cause distorted bands. We can rectify this error by cooling. These distorted bands can appear due to higher salt concentrations, bubbles in the gel, insoluble materials in the gel and irregular polymerization of gels. In some cases, the protein bands can be seen in the unloaded tracks. This is quite common in the case of slab gels. This is due to various factors like a contaminated running buffer and poor polymerization. There will be incomplete teeth formation. This will result in the leakage between the sample wells. The overloading of wells result in the sample overflow. We observe unnecessary bands in the tracks occasionally. This can be due to buffer contamination. The band will be seen in almost all the loaded tracks. The unexpected bands can be due to sample overflow. In some cases, we observe less resolution of the bands. The less resolution is due to less polymerized gels. Applying inappropriate pH values for buffers and sample overloading can diminish the resolution. If the gels are stored for longer periods, there will be a diffusion of buffers. This diffusion can occur between stacking and resolving gels. In some cases, we can observe background staining following the individual tracks. This can be due to the overloading of the stains. If the sample buffer is with impurities, we can expect high background staining. Occasionally, we observe protein streaking on the gel. This can occur due to the precipitation of proteins. There will be dissolution whilst electrophoresis. It is important to centrifuge post protein denaturation. This step will eliminate the insoluble materials. It is always better to reduce the sample volumes to avoid leakage. If the proteins aggregate, they will reach the stacking gel. This problem can be solved by employing a continuous buffer system. With this buffer system, the protein aggregation in the sample will be prevented. In some cases, there will be the formation of double bands or protein dimers. The formation of double bands is due to the use of aged buffers.

We observe protein smears in the gels. This can be due to the omission of SDS in the buffer solution. There will be a diminished resolution. This can be due to minimal dissociation of proteins from SDS. There will be an irregular run time whilst electrophoretic run. In the case of long run times, the running buffer can be diluted. In some cases, the gel buffer can be highly concentrated. In the case of a shorter run time, the running buffer can be concentrated.

The choice of the buffer can be important. Through several experiments, it was observed that proteins can have different stability, solubility and charge with varying pH values. They are different cofactor demands, ionic strengths and ionic species. It is an important task to select the correct buffer in the gel systems. The buffer to be chosen must follow certain conditions set by the researcher prior to the electrophoretic run. The proteins present in the buffer must remain soluble and stable. The proteins should be charged during the run and buffer must provide this condition so as to facilitate the perfect electrophoretic run. If the pH of the buffer is increased from the isoelectric points, then we can say that there is more charge on the proteins. Due to this, we can expect reduced diffusion of bands. If the pH is close to the isoelectric points, there will be large charge differences and so we can get clear separation. We can use continuous and discontinuous gel systems in the electrophoresis. It is to be kept in mind that the discontinuous buffer systems can provide us with high resolution. There are about 4000 buffer systems (Chrambach and Jovin et al., 1964). Low pH system resolving proteins occurring at pH 3.8 and high pH systems resolving proteins at 9.5 can be considered as the most popular buffer systems (Reisfeld et al., 1967). There are some proteins that aggregate. Due to protein aggregation, there will be precipitation. There will be a streaking problem.

The sample buffer should contain 10% glycerol or we can use sucrose. It is important to use appropriate tracking dye like bromophenol blue in case of high pH systems. We can use dyes such as methylene blue, methylene green and Pyronin Y in case of low pH systems. We can use ¼ diluted buffer stock to load the proteins onto the wells in the gel. We can add 1 vol of the buffer with a sample and load directly on the gel. In the case of low pH buffer systems, we need to connect the upper electrode to the anode of the power source. The preparation of protein samples must be done at 4°C. If the temperatures are low, we expect lesser problems with the proteases.

The gels may get heated due to continuous power supply and this will cause protein denaturation. We can reduce the current or we can perform the electrophoretic process in a cold room so to avoid these problems. If we don't know the native properties of the proteins and their isoelectric points, then the choice of buffer to be used can become difficult. In order to reduce this complication, Schagger and Jagow (1991) have developed the Blue native-PAGE method. This method is called a charge-shift method. Coomassie blue G-250 can be added to proteins and so the charge will be induced onto the proteins. We can assume that there will be a charge shift. This dye is negatively charged and it mostly binds to the hydrophobic portions of the proteins. With this charge shift, we can conveniently separate the proteins. In this technique, the Coomassie dye will bind on the protein surface and in the case of SDS-PAGE, the binding will be with the interior sides. The binding capacity in this technique will be variable. Some of the advantages of BN-PAGE are enhanced solubility and reduced protein aggregation. There will be reduced protein denaturation at 7.5 pH. In this technique, we can stain the native proteins so that they can be visible on the electrophoretic gel.

We can use both SDS-PAGE and BN-PAGE to determine the nature of the proteins in their oligomeric states. This technique is more sensitive and with this technique, we know the homogeneity of the proteins in their native states, oligomeric states and even we can analyze the molecular mass. Acid-urea PAGE can be considered as an important technique in the protein purification process. We can resolve histone proteins based on the molecular weights using the SDS-PAGE system. This system has certain problems. With the SDS-PAGE system, we may not separate the histone variants. The modified forms of the histone proteins may not be resolved with this technique. The modified forms can be formed due to post-translational modifications like phosphorylation, methylation and acetylation. The main problem while using SDS-PAGE to resolved these modified forms is that they share the same molecular weights. In order to separate such forms with similar molecular weights, it is important to search for other properties such as charge. The acid-urea PAGE can be used in the histone analysis and we can use the same technique to separate other proteins (Harwig et al, 1993). Generally, urea will be applied in the protein denaturation process. Due to the addition of urea, the proteins will be more soluble and so the mobility will be decreased. This can be attributed to the unfolding of the proteins. We can use Triton X-100 to enhance resolution. The histone proteins can get attached to the Triton molecules. Post addition of Triton, the migration will slow down. The binding of detergents can be lowered by urea. The main advantages of this system are that there will be reduced background post silver staining. There is no need for pre-running. We can expect speed polymerization rates.

In the case of transverse gradient gel electrophoresis, there will be slab gel which can be classified as continuous and exhibit linear gradients. This gradient will be perpendicular to the protein migration tracks. The gradient will be of acrylamide. In this technique, the proteins can be applied on the top side of the gel. The mobility of the proteins is based on their size and charge. The confirmation of proteins is crucial in determining mobility. The conformation changes in the protein can be identified through electrophoresis. Transverse temperature gradient gel electrophoresis or TGGE can be classified as a transverse gradient technique. We can measure the transition temperature of cold and heat- induced conformational changes present in the proteins. We can even detect conformational variations.

Tricine SDS-PAGE can be an important technique in resolving the proteins. In SDS-PAGE, it is difficult to resolve oligopeptides and proteins with 14 kDa molecular weight. In Tris SDS-PAGE, we can eliminate the use of urea and glycine. The detergents can be divided according to charge. We can classify them into cationic, anionic and neutral detergents. There are many types of detergents that can be used in the electrophoretic process. Mostly, SDS can produce crystals if the temperatures get drop and, in that case, we can use lithium dodecyl sulfate. SDS may cause aggregation of proteins in whilst electrophoretic run.

3

Protein Chromatography

CONTENTS

3.1 INTRODUCTION

The isolation of proteins can be classified as a key objective of protein purification techniques. The proteins must be devoid of biomolecules that are not required in the process. The conventional chromatographic techniques or protocols are based on the binding affinity, physicochemical properties and size of the proteins. There are many advances in bioinformatics and molecular biology that led to the development of proteomic platforms. The knowledge of proteins and functions they can perform is important in the protein purification schemes. The basic biological process must be studied. The protein purification can be defined as a set of procedures used to isolate desired proteins from a complex biological system. We need to frame a perfect strategy of purification before making any studies related to the structure and function of the proteins. The proteins are made up of amino acids and they will be arranged in long chains. The amino acids are linked via peptide bonds. The peptide bond is formed between the amino group and the carboxyl group. The first amino acid in a protein can be free so as the end amino acid. This is true for polypeptides too. The amino group of the first amino acid and the carboxyl group of the last amino acid will be free or unbound. We call the start of the polypeptide as N terminus and end as C terminus.

There are about twenty naturally occurring amino acids that can be differentiated with the variable R group. The R group can define the physicochemical properties of any amino acid. The conformation of the proteins can be based on their amino acids sequence. The proteins can be complex as the amino acids will push the proteins to attain certain folding along with other bonds (Sun et al, 2004). The salt bridges, hydrogen bonds, hydrophobic packing and van der Waals interactions can keep the tertiary structure of the proteins intact (Kang et al., (2005), Bartlett et al (1976) and Joseph et al., 1997). In the biological systems, there will be several protein modifications post -translation process. Some of the post -translational modifications include chemical alteration of amino acids. This can be done by processes such as elimination, deamidation and deamination. There will be a covalent linkage between peptides and this linkage can be maintained or initiated by ISG15, ubiquitin and SUMO protein. There will be certain structural modifications. Most of the modifications are brought about by racemization of proline amino acid, the formation of disulfide bonds and proteolytic cleavage. Proteins can be classified as fragile and complex. Anyone interested in knowing the structure of the protein must purify it so that it becomes contaminant free. In 1790, Antoine Fourcroy tried to isolate certain substances from the plant extracts which were having the same characteristics of egg albumin. Franz Hofmeister produced protein crystals of ovalbumin. John Kendrew and Max Perutz (1957) described the molecular structure of the protein with the help of X-ray crystallography. They described myoglobin and hemoglobin (Strandberg, 2009). During World War II, several attempts were made to purify proteins present in the blood. They used the Cohen fractionation process to purify albumin (Cohen et al, 1994).

The charge of the proteins can be based on the total charge of the amino acids that constitute the protein structure. The charge on the proteins can be different and it is sensitive to pH. The charge of the protein at an isoelectric point will be neutral. The solubility of proteins in water can be due to interactions between the hydrophilic amino acids on the peptide chains and water molecules. The solubility of the proteins can be changed by altering the ionic strength, temperature and pH. Biological affinity can be used in the protein purification process. The antigen-antibody interactions, ligand and receptor and enzyme -substrate interactions can be used in the chromatographic techniques. The antibody- dependent chromatography is being used to purify the desired proteins and, in some cases, we can even purify the antibodies.

Proteome was first introduced by Keith Williams and Marc Wilkins in 1994. They used this word to describe the entire proteins that can be expressed by a genome. Sophisticated techniques such as ESI or Electron spray Ionization and MALDI or Matrix Assisted Laser Desorption/ Ionization have helped the protein analysis (Wang et al, 2012). The 'Bottom up' method of identification of proteins involves protein digestion before mass spectroscopy studies (Whittaker, 2013).

3.2 GEL-FILTRATION CHROMATOGRAPHY

This technique can be classified as a robust method that allow us to separate high yields of biomolecules. This method can be considered as a variant of partition chromatography. This technique can be used to resolve molecules based on molecular sizes. This technique can be called as molecular sieve chromatography or size exclusion chromatography or gel-exclusion chromatography or gel permeation chromatography. In this technique, the molecules will be resolved or partitioned between the stationary phase and mobile phase based on their respective sizes. We can use bead form and it comprises of 2 liquid volumes. The external volumes comprise liquid which can pass through beads. The internal volume comprises of liquids present inside the beads. The external volume can be called as void volume.

There are many factors that are important in the gel-filtration process or system. They are storage and cleaning of the column, flow rate, eluent type, column parameters, size and concentration of sample and matrix type. The matrices being used must be inert so that there will not be adsorption on the matrix (Maurer et al, 1978). The matrices used in the materials are dextrans, agarose, polyacrylamide, silica, cellulose, polyvinylethylcarbitol and conjugated materials. The matrices that are used in the gel filtration process can be made up of cross-linked polyacrylamide.

In the size exclusion chromatography, we can use hybrid inorganic and organic particles (Huang et al, 2003). In order to reduce the nonspecific ionic reactions, we can use a modified surface with diol groups. We can use BEH particles or bridged ethyl hybrid particles. The matrices made from agarose can be tolerant to the alkali materials. So, these matrices should be cleaned with NaOH or sodium hydroxide. This can be classified as an effective technique to enhance the life of the matrix. Silica can be used as a support material. The silica gel can be prepared with sodium silicate. It is a porous, vitreous and granular mineral. It can be processed into a beaded or granular form. The average pore size can be about 2.4nm and it will attract water. Silica can be used because of its narrow distribution of pore size and large pore volume. We can expect a high resolution and sharp peaks in most of the cases. The silica particles can be fragile and brittle in nature. They can break at high pressures. We can exhibit higher resolution in the case of gel-filtration chromatography. The sample must be applied in small volumes. We can expect about 1% of the total bed volume. The gel filtration chromatography can have minimal handling capacity. The sample concentration can be regulated by the viscosity. The viscosity can get increased with the sample concentration. Due to high viscosity rates, we can expect irregular migration of the samples through the column. At any time, we should not use the sample above 20 mg/mL. We can expect high resolution with long columns. This technique can separate the molecules based on their sizes. We can use any type of eluent. The elution conditions like protease inhibitors, cofactors, essential ions and pH must be selected according to the properties exhibited by the desired proteins. The ionic strength of the eluent must be higher so that the protein-protein associations and protein-matrix associations can be minimized. These associations can be managed by the van der Waals and electrostatic interactions.

Lower flow rates can produce high resolution because the resolution can flow rates are inversely proportional. The matrices used in the gel filtration process can be cleaned with non-ionic detergents or 0.2M sodium hydroxide. The matrices can be stored in dark at 4°C. The silica-based columns can have large volumes and this will help us to attain high resolution. The hydroxylated meth-acrylic polymer resins can be used in SEC media. This media available in the market with diverse particle diameters and pore sizes. Proteolysis can be considered as an important problem with this technique. The target proteins serve as substrates needed for the proteases to act on. We can observe reduced recovery in this case. We need large effluent volumes that can relatively increase the running costs. This technique has limitations with the sample handling quotas. Despite limitations, the gel filtration chromatography was used in purifying many peptides and proteins obtained from diverse sources.

The gel filtration can be used in the purification of alpha-lactalbumin and beta-lactoglobulin (Rojas et al, 2011). There are several protein engineering techniques that are used in nanotechnology (Goodsell, 2013). Proteins can bind to polyethylene glycol. This reduces antigenicity and shelf life. The gel filtration chromatography is being used to separate DNA, RNA and transfer RNA. We can even separate adenine, guanine, uracil, thymine and cytosine with gel filtration chromatography. We can separate circular M13 DNA and DNA of linear phage lambda (Gonzalez et al, 1990). We can isolate plasmid DNA by gel filtration (Kondo T et al, 1991). The size-exclusion chromatography was used to recover the HbcAg or hepatitis B core antigen (Ho et al, 2009). It is important to know that the endotoxins are dangerous in the bloodstream as they may induce fever. According to GMP or Good manufacturing practice protocols, it is important to remove the endotoxins. There are several end products that can be produced via non-GMP protocols that are preclinical biologicals. The size exclusion chromatography can be joined with a dynamic light scattering instrument to produce absolute size exclusion chromatography or ASEC. The dynamic light scattering is a technique that can use light scattering patterns. This technique can be called as quasi-elastic light scattering or photon correlation spectroscopy. In this technique, the laser can be treated as a light source. This technique can be used to know the distribution of the size of the particles in the polymer solution or in suspension (Berne et al, 1999). This technique can be used to detect complex fluid behavior in polymeric solutions. We can measure the size of the molecules once these molecules flow through the DLS system. The flow will start from the size exclusion column. The hydrodynamic size of the molecules will be determined. The molecular weights of these molecules will be ignored. We can calculate molecular weight via hydrodynamic properties by Mark-Houwink calculation (Wang et al, 1991).

The gel-filtration chromatography can be classified as an appropriate alternative for SDS-PAGE. This technique can be used to know the relative molecular masses. It is important to know that the volume of the globular protein is linked with the molecular weight logarithmically (Andrew, 1965).

3.3 IMMUNOAFFINITY CHROMATOGRAPHY

The immunoaffinity chromatography or IAC can be classified as an antibody-based separation technique. This technique is a powerful isolation and purification technique. The antibodies that are produced from this technique can be useful in several fields starting from environmental monitoring to clinical diagnosis. This technique uses the binding efficiency of an antibody and ligand.

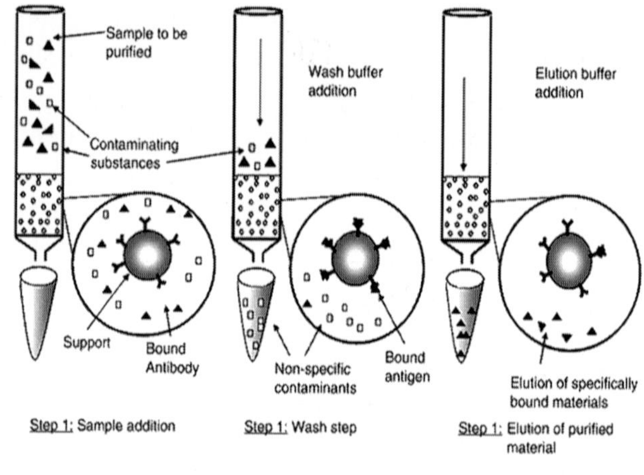

Figure 3.1. Immunoaffinity chromatography.

Source: https://link.springer.com/protocol/10.1007/978-1-4939-6412-3_3

The critical parameters should be optimized so that this technique can be flawless. We need to consider effective conditions for elution, immobilization techniques, matrix support and biological affinity pair. This technique is based on the recognition between the complementary ligand and antibody molecule. This technique has many advantages than conventional techniques. There will be molecular selectivity due to high affinities between the antigens and antibodies (Delaunay et al, 1995). In this technique, spatial complementarity will be established and this can be due to the addition of all the intramolecular bindings. This technique can be used in biological laboratories and pharmaceuticals. We can use this technique for separating, isolating and purifying proteins. Antibodies can be classified as soluble granular proteins that are generated by the immune system. The antibodies are produced in response to the antigen intrusion. There are about 10^{11} antibodies that have different specificities (Murphy et al, 1996). Most of the antibodies are present on the surface of the B cells and so they are classified as membrane- bound forms. They act as receptors for antigens. These antibodies are soluble when they are present in lymph or blood. These antibodies do participate in the humoral response against the bacterial cells.

The structure of an antibody is made of two identical light chains. These chains are about 25 kDa. There are two heavy chains which can be about 50 kDa. The heavy chains and light chains will be connected by disulfide bridges and noncovalent bonds. (Figure 3.2)

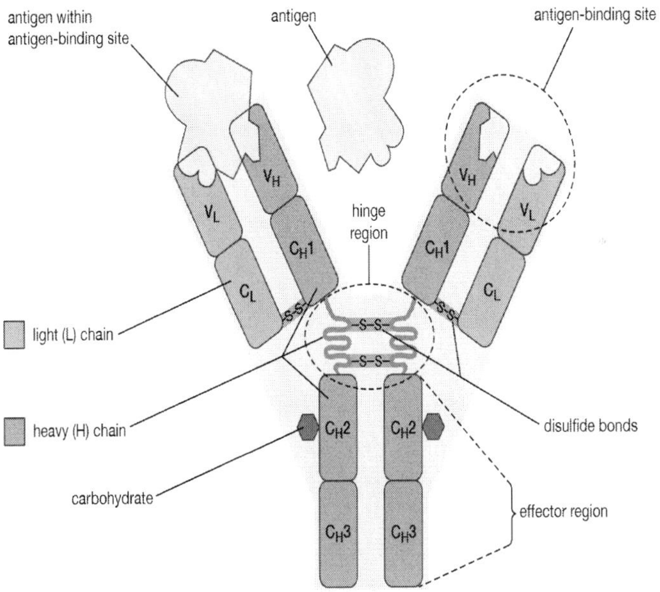

Figure 3.2. Heavy and Light chains.

Source: https://www.britannica.com/science/antibody

The specificity of the antibodies will be based on the hypervariable regions. These regions have residues that are complementary to the antigens and the structure is also complementary. The residues are called as CDRs or complementarity determining regions. The composite surface is formed due to juxtaposed variable light chains and heavy chains at the end of the antibody (Barbas et al, 2005). We can observe five antibody types. They are IgA, IgD, IgE, IgG and IgM (Figure 3.3). The concentration of IgG in the serum is about 13.5 mg/mL (Shantha, 2003). In the case of immunoaffinity chromatography, we may use G class immunoglobulins. The antibodies can be raised against antigens. There are many methods that can be used in the production of antibodies.

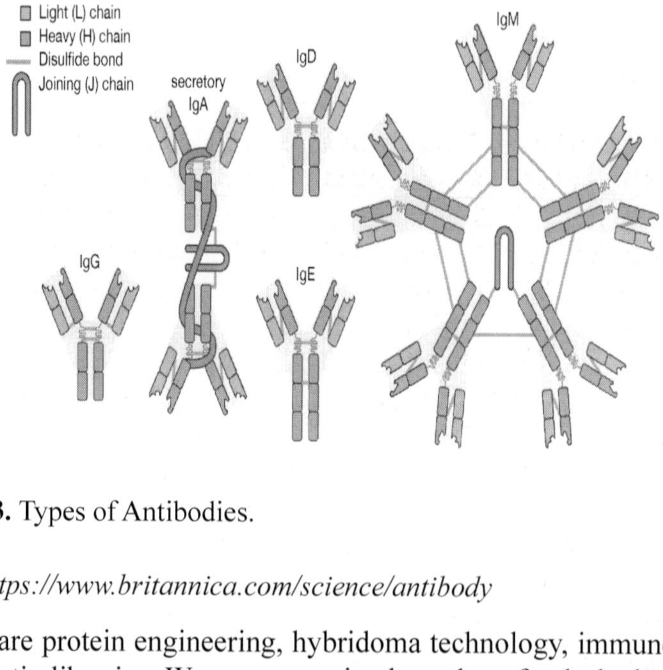

Figure 3.3. Types of Antibodies.

Source: https://www.britannica.com/science/antibody

They are protein engineering, hybridoma technology, immune libraries and synthetic libraries. We can use animals such as fowl, donkeys, sheep, goats, guinea pigs and rabbits to generate antibodies in the polyclonal serum. We can get the polyclonal antibodies from the serum of the animals previously injected with antigens. In this serum, we will get antibodies with varying specificities and affinities (Fitzpatrick, 1981). Kohler and Milstein in 1975 have explained the production of monoclonal antibodies. They used the somatic cell fusion process between the immune lymphoblast and myeloma (Kohler and Milstein, 1982). The mono-specific antibodies were produced using hybridoma technology (Figure 3.4). The monoclonal antibodies can be used in food analysis, environmental analysis and immunodiagnostics. They can be used in treating infectious diseases, cancer, inflammation and autoimmune diseases (Roque et al, 2007).

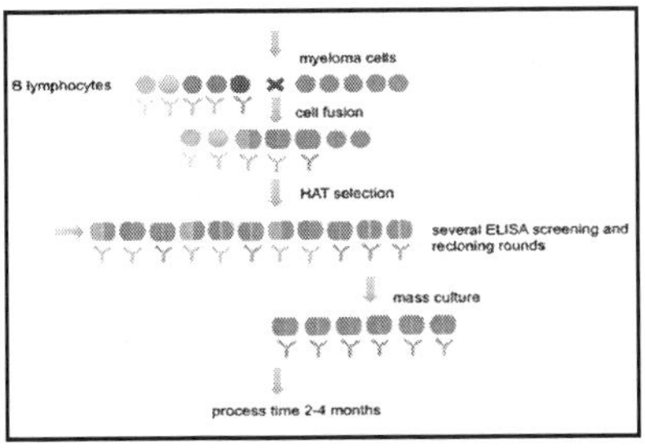

Figure 3.4. Hybridoma Technology.

Source: https://www.nature.com/articles/s41598-020-58571-w/figures/1

The recombinant technology has helped us to produce affinity ligands. These affinity ligands are cheaper and can be used in the immunoaffinity chromatography. Phage display is used to present the protein and peptide libraries. There are many conditions to be followed whilst Immunoaffinity chromatography or IAC. There are three factors that reduce the applicability of IAC. The antibodies and antigens are highly specific and this will produce problems in the elution process (Muronetz and Korpela, 2003). The antibodies can be immobilized on the matrix and so the Ag-Ab capabilities will be reduced (Turkova, 1999). The antibodies can be tissue- specific and species- specific. It is important to isolate specific antibodies. We can use proteins which are immobilized on the bacterial surface. The synthetic ligands can be used in a variety of applications due to their low molecular weight. The synthetic ligands have the same selectivity as the natural ligands. They are cost-effective, durable and exhibit high binding capacity. The synthetic ligands can be resistant to both biological and chemical degradation. Mimetic peptides are classified as synthetic peptides. These peptides can be better than conventional ligands. Pseudobispecific ligands can be cheaper and considered as an alternative to the natural ligands. They have better absorption capabilities and immobilization capabilities.

Matrix is an important component that can be used in this chromatography. Without a matrix, the ligand may not function. The conventional matrices can be used with minimal pressure resistance and we rely on gravity flow. This process can be facilitated by peristaltic flow or vacuum flow. Some of the matrices used in this technique are cellulose, dextrose and agarose.

There are some synthetic polymers like zirconia, silica, polyethersulfone, polymethacrylate and acrylamide that can be used in this chromatography. These supports can be used in immobilization of hormones, nucleic acids, lectins, antibodies and enzymes. There are certain limitations to these matrices. The stability and flow rates at high pressures is a problem. The chemical properties, hydrodynamics and slow mass transfer are some of the issues with these matrices. We can use azlactone beads, glass and derivatized silica to get a high performance support. Due to the superior qualities, we can use these supports in HPLC methods. The hydrophilic matrix can be used to reduce the antibody leaching and nonspecific adsorption. The immobilization of antigens and antibodies to affinity support matrices should follow certain conditions. They should have large pore sizes and the particle size should be uniform. They should be mechanically stable and must be activated easily. They should be biologically inert and chemically non-reactive.

The elution stage in this chromatography needs disorientation of the link between the ligand and receptor. This results in a lowered binding affinity so that the ligand may return into the mobile phase (Firer, 2001). The elution conditions must be modified according to the purification process. The elution buffer must break bonds that maintain the protein to protein interaction. A simple buffer can be selected based on the expected functionality and convenience. The analyte elution is based on the water-organic mixtures, pH differences, rise in temperatures, chaotropic agents and displacer agents. The displacer agents can be used as competitors to the ligands. Some of the elution methods can use high pH buffers. These buffers can break the ionic bonds. We can use triethylamine. We can use acidic pH buffers like glycine- HCL. Singh et al. (2009) have described 20 elution methods. We can use organic solvents like methanol in IAC. We can use chaotropic agents as elution agents. These chemicals changes the water structure at the point of affinity. Histidine, glycine, lysine and proline can be used as elution buffers. The antibodies can be classified based on biochemical and physicochemical properties. The procedures associated with characterization can give us information on the functionality, purity, sensitivity, affinity and integrity of antibodies. Some of the qualitative approaches are isoelectric focusing, Western blotting and SDS-PAGE. The quantitative approaches are bicinchoninic acid assay, Lowry and Bradford assays. The antibody concentration can be estimated using ELISA or Enzyme Linked Immunosorbent Assay. Most of the ELISAs can be considered as the foundation of several immunoassays. This technique can be used to estimate the cross-reactivity, specificity, affinity, sensitivity and concentration of antibodies. In Mass spectrophotometry technique, there will be ionization of samples. Due to this ionization process, there will be a charge on every compound. This property can be used to calculate the charge to mass ratio. Mass spectrophotometry or MS (Figure 3.5) is classified as an important tool in protein characterization. We can get a high resolution (Zhang et al, 2009). The applications of IAC are immense and diverse. This technique can be used from environmental monitoring to clinical therapeutics. IAC can be used as a separation tool, detection, extraction and purification tool and it is based on the purpose of the study. We can remove certain solutes from a sample and IAC can also work as an analytical method.

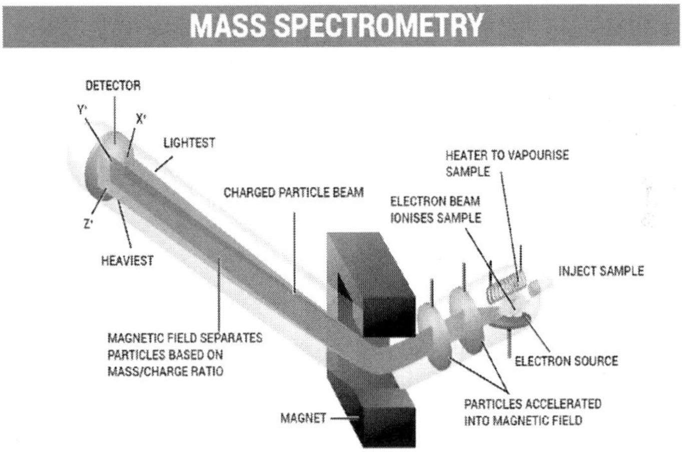

Figure 3.5. Mass Spectroscopy.

Source: https://microbenotes.com/mass-spectrometry-ms-principle-working-instrumentation-steps-applications/

We can couple mass spectroscopy (MS) with IAC to produce a combined system that is more effective and accurate. This technique offer us high selectivity whilst Ag-Ab interactions. We can avoid reduced ionization efficiency with volatile buffer salts such as ammonium acetate. We can enhance the sensitivity by adding acetonitrile and methanol to the eluant. This must be done before ionization (Hage, 2018). IAC can be used in both the detection and extraction of contaminants in foods. With this technique, we can detect and extract mycotoxins from the foods. Some of the mycotoxins that can be extracted through these systems are fumonisins, zearalenone, ochratoxins, and aflatoxins. We can isolate contaminants like sulphonamide and quinolones. The herbicide like atrazine can be extracted from foods and soils (Chuang, 2007). We can use this technique in detecting alkylbenzenesulfonates and microcystins in freshwater.

We can detect tetrodotoxin in intoxicated patients and estrogen in pregnant women using this technique. We can identify biomarkers that serve as targets in the biological fluids. We can detect the anti-nicotine metabolites. Some of the therapeutic applications of IAC start from the isolation of monoclonal antibodies to recombinant proteins. It can be used as a regular procedure for both purification and detection.

3.4 PROTEOLYSIS

The stability of the proteins can be defined by the conserved biological function and molecular integrity even in adverse conditions and influences such as deleterious conditions or heat (O' Fagain, 1997). The deleterious conditions can be defined by the proteolytic enzymes in the protein chromatography technique. The proteolytic enzymes can be called as proteases. Proteolysis is nothing but the protein degradation in the cell environment and it is defined by a balance in the rate of degradation and rate of synthesis to maintain homeostasis. There are considered as nature's swiss army knife (Alnemri, 1997). The living cells use proteases to regulate protein turnover. It is important to a cell to catabolize proteins continuously so that there will be free amino acids in the cytoplasm. The proteases are less dominant in case the purification is for recombinant proteins. The recombinant proteins can be produced in *E.coli*.

3.5 DESIGN

The design of the chromatographic process can be constrained by the economic viability, robustness and quality of the product. The reproducibility and robustness are considered as critical in defining the safety of the product. The product quality is important in the production of biopharmaceuticals. The economic viability depends on the depreciation, labor, process chemicals, solvents, process productivity and yield. The productivity rate of the chromatographic process can be based on the quanta of the product process per unit volume and time. In several applications, the cost of the stationary phase can be considered as an important factor. It is important to enhance the productivity rate so that the overall process can be beneficial. The productivity rate will be based on the selected operational conditions and the stationary phase. It is important to establish perfect choices for chromatographic processes. The utilization of chemicals, solvents and labor can be crucial in selecting the choices. The buffer preparation and disposal of the buffers can be considered as important components which need to be considered.

The packing of column and unpacking must be important as it comprises of labor requirements. It is based on the design of the column hardware and stationary phase. We need to consider the depreciation of the equipment that is fixed. The costs needed for equipment must be minimized or reduced so that the operational costs can be controlled. We can use simulated moving bed systems so as to achieve a high rate of productivity. In these systems, there will be a limited consumption of chemicals. The overall chromatographic process is dependent on economic, regulatory and technical factors. In a biopharmaceutical industry, three major steps are to be considered are capture, separation and polishing. Each step has a different set of goals. In the capture step, there will be a concentration of the product mainly on the stationary phase. This can be done by removing the impurities. The impurities are protein contaminants, endotoxins, DNA and other small molecules. The properties of these impurities can be different from the target products. The impurities must be important so that we can expect high yields. It is important to know about the impurities in high resolution. This resolution can form a second goal in the capture step. In the capture step, there will be a stationary phase that exhibit high binding capacities. We can use shorter columns with high flow rates. The stationary phase may contain larger particles. We can use selective affinity adsorbents. Protein A can be used to capture IgG. In the capture step, we can use steep gradients and step elution to recover the products. Most of the other steps are needed to resolve impurities that are similar to the target products. Some proteins that are impurities in the current purification process can have the same pI and hydrophobicity or isoforms. In the case of shallow gradient operations, we may need longer columns and lower flow rates. We need to use the polishing step to remove the process chemicals. In some cases, the impurities will be produced whilst the process itself. This includes aggregates and dimers. We can use size exclusion chromatography for this purpose. This chromatography has minimal capacity. It can be used to eliminate the aggregates and we can keep these eliminated aggregates in the buffer. The concentration of the products can be high in this process. Size Exclusion chromatography has less capacity and the economics may not be altered much.

There are certain intermediate steps that are important in designing the downstream processing. Mostly, the downstream processing will utilize the orthogonal stationary phase. In this step, we use cation exchange and anion exchange. The dilution process will lower the protein concentration and so there will be enhanced feed. The batch capture process can be designed so that the target proteins can be separated and the impurities will be separated from the contaminants.

The time needed for the wash step may be estimated from the unbound contaminants with a linear isotherm. The elution time can be based on the isotherm parameters. It is important to use Clean-in-place protocols or CIP protocols to remove the impurities with string binding affinities. These protocols can be used in case of clipped functions, unfolded proteins and aggregates. It is important to perform validation studies so that the concentration of the cleaning agent and the time needed for this agent to clean can be known. It is important to know that the flow rates are lowered. The design of the column is based on the nature of the adsorbent particles. These particles can be compressible or rigid.

3.6 GRADIENT ELUTION CHROMATOGRAPHY

The gradient elution chromatography (Figure 3.6) is used frequently in the protein chromatography process as it is beneficial. The mixtures used in this technique can be complex and we can expect wider adsorption strengths.

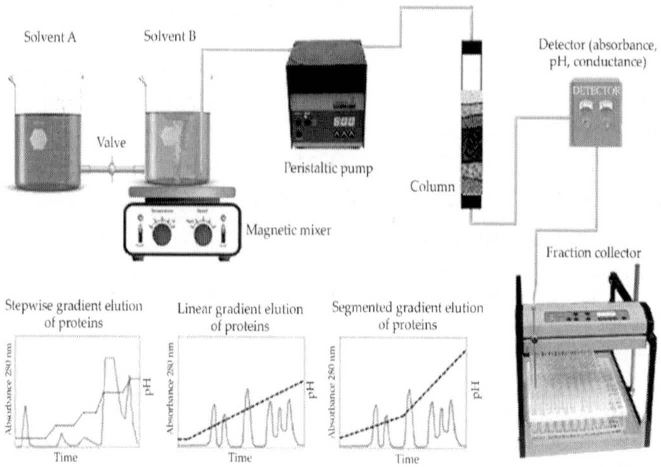

Figure 3.6. Gradient Elution Chromatography.

Source: https://www.fractioncollector.info/fraction-collector-protein-purification-chromatography/

There will be some components that can be partially bound and they will be in the mobile phase. The gradient elution chromatography can be extremely sensitive in nature. The gradient to be used is based on the chemical interaction. The gradients which are having enhancing organic modifier can be used in case of heavy salt concentrations. The gradient shape can be variable. Linear gradients that are present at the entrance of the column can be linear. In this chromatography, the feed mixture will be loaded onto the column. This loading will be done by the injector system or load pump. The composition of the mobile phase will be changed gradually according to the slope.

If the concentration of the modifier is increased linearly over a certain time period, then we will get linear gradient elution. The modifier gradient can be used in several practical applications. Mostly the modifier gradient will be delayed in the column. There are two gradients, one is temporal gradient and other is axial gradient. The temporal gradient can be produced at the entrance of the column. The axial gradient can start from inlet to outlet and it is available at every time point. The compression of the band is due to the axial gradient. Due to this, there will be peaks. This can also occur due to isocratic elution. The feed components and modifiers can be explained by linear isotherms. The modifier retention may not be controlled by the components of the feed. In the case of excess modifiers, the adsorbent surface can be saturated by the modifiers. These conditions are seen in protein chromatography.

3.7 COLUMNS AND EQUIPMENT

The flow rate and pressure can define the design and elements of any chromatographic equipment. The pressure can be due to the length of the column, flow rate and particle size. In general, the pressure will be determined by the application type. In the case of high-resolution techniques, it is important to use more plates. Long columns are needed in the case of high-pressure systems. We can observe two types of velocities in the case of column chromatography. They are superficial velocity and interstitial velocity. Mostly, the superficial velocity will be based on the volumetric flow rate. Latest chromatographic systems can be used at pilot-scale and laboratory scales. We need two pumps in the mixer. Feed application system or sample application systems must be installed. We need separation columns ad detectors. It is important to have a device to collect fractions. Computer- aided regulatory systems and electron data acquisition systems can be used. In the case of isocratic elution, a mixer may not be needed and it holds good even for stepwise elution. If buffers are mixed with the stocks, it is important to use pumps. A sample or feed application system can be a valve that is linked with a calibrated loop. We can even use a feed pump for that matter. The separation column can be about 0.3ml. In the case of industries, we can use a 1000 liter column. We can use one or more detectors and the number is based on the type of application to be used. We can use valves to collect the fractions. We can use pumps and valves and they are regulated by a computer control system. It is important to have certified equipment in the case of electronic data acquisition systems. Most of the components will be integrated with the computer systems and this will help us to program the runs as and when needed. We can use the computer-based programs so as to run the chromatography with a diverse range of columns and conditions. The long- stroke piston pumps include syringe pumps in sync mode and syringe pumps. The former can be used in FPLC and the latter can be used in a laboratory and with good precision. The reciprocating pumps include dual- piston and single- piston. Both these pistons can be used in the laboratory with a high rate of precision. The Fast fill single piston can be used in the lab with high precision. The peristaltic pumps can be used in the lab at large scales and we may not use these pumps in case the pressures are well beyond 3 bar. The centrifugal pumps may not be suitable. The rotary lobe pumps can be used in the lab for large scale applications and it reaches up to 400m^3 per hour. The diaphragm pumps can be used in large scale applications up to 1000m^3 per hour. The syringe pump can have check valves but there will be no pulsatile flow. The long-stroke piston pump in sync may not have both check valves and pulsatile pumps. The check valves and pulsatile flow can be present in the single and dual -type pistons. The check valves can be present in the fast fill single piston but the pulsatile pumps are absent. In the case of peristaltic pumps, there will be pulsatile pumps and the check valves will be absent. Both check valves and pulsatile pumps will be present in the diaphragm pumps. Both of them are absent in the centrifugal pumps and rotary lobe pumps.The positive displacement pumps can be used to regulate the flow even at high pressure situations. The syringe pumps can be very accurate and we can expect non-pulsatile flow. This pump alone can be enough to run the entire chromatography unit. We can use these pumps in capillary chromatography.

The dual long stroke piston pumps can be used in the synchronized mode and can be used in fast Protein Liquid Chromatography or FPLC. In FPLC, the liquid will be delivered by one piston and the second piston will refill it. The primary advantage of the syringe pumps is that the check valves are absent. In this case, the pressure wave will be produced whilst valve switching. We can use reciprocating piston pumps in High Performance Liquid Chromatography or HPLC. We can use them in the form of tandem units or single units. Most of these pumps can produce pulsatile flow. The pumps need the check valves and we can expect fouling due to the continuous flow of the biological fluids. They can produce the pulsatile flow and may need pulse damper. In modern labs, we need to use tandem piston pumps. We can use these dampers to make the flow pulsations smoother and lower the pressure spikes. The pulse dampers can have an elastic tube that is flat. The oscillation velocity can be compensated. The diaphragm dampers can have liquid or air-filled cushions. This can be compressed by high pressures. The air solubility in the aqueous media can get lowered with decreasing temperatures. There will be fluctuations in temperature. This is due to heat produced by solution and ambient changes. The air can be produced in the mobile phase and hence there will be bubble formation. If there is bubble formation, we can expect cracks in the bed and non-uniform liquid distribution over the column. It is important to perform a degassing process by vacuum degassers. This will allow for uniform air distribution.

The buffer mixers can be critical in the chromatographic systems. They can be used in the production of the gradients. There are 2 buffers that can be used in different proportions so as to produce gradients. We can use either high pressure mixers or low-pressure mixers for this purpose. In the case of low-pressure systems, there will be a single pump with inlet valves. In the case of high-pressure systems, there will be independent pumps. The flow rates will be adjusted accordingly. It is important to mix thoroughly so that there will be accurate gradients. There will be a small tank and the contents will be mixed with a magnetic stirrer. The constant upflow will prevent bubble formation. The higher viscosities and increased flow rate should be ensured so that there will be effective mixing. Static mixers and impingers can be used in swift mixing. The connectors, tubing, detector and mixers can be used in column broadening and it is critical whilst data analysis. The process columns will be large and they are made up of plastics, glass and stainless steel. The diameter of these columns will be about 2m and they can support volumes upto 1000 liters. There are several requirements whilst designing the process columns. The sanitary design can permit the cleaning of columns. The flow can be distributed starting from the inlet pipe. The inlet pipe will be a few centimeters in diameter. We need to know the tendency of the distribution of the mobile phase in an uninterrupted manner over the cross -section of the bed during the runs. The filling ability of the column in the stationary phase whilst packing must be considered.

The particles used in the chromatography run can be present in the column by a sintered frit or mesh. They are produced from metals or polymeric materials. We can use multilayer designs. The pore diameter or the size of the mesh can be based on the size of the particles present in the stationary phase.

4

High Performance Liquid Chromatography

CONTENTS

4.1 INTRODUCTION

High performance liquid Chromatography or HPLC (Figure 4.1) can be classified as an important technique in the analysis and purification of proteins. This technique has played a vital role in the biomedical sciences and biological sciences. This technique is a success that can be attributed to the high recoveries, ease of selectivity and reproducibility. We can get high resolution even though the impurities are sharing the same properties as the sample.

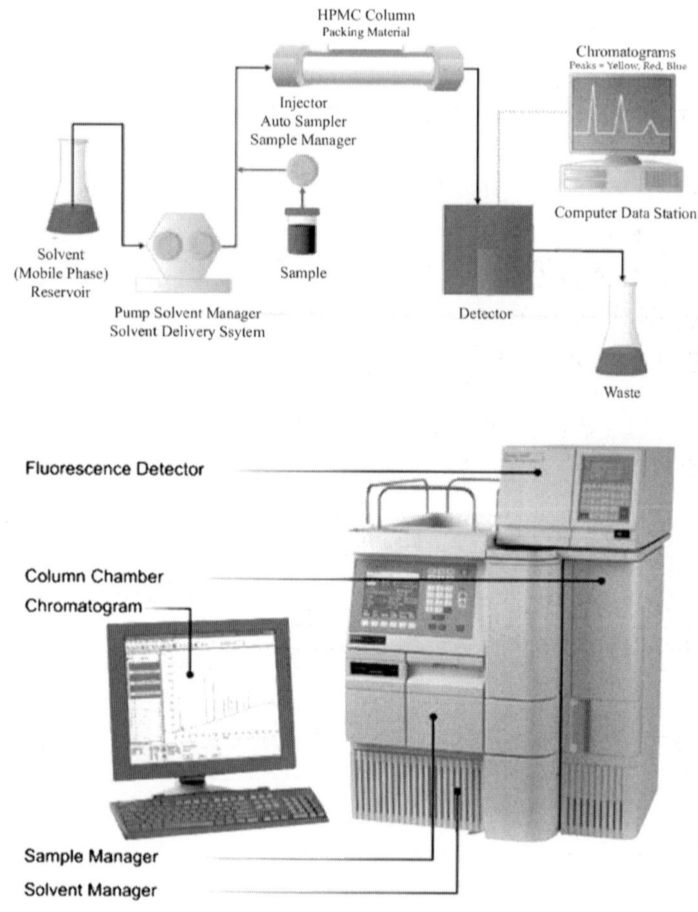

Figure 4.1. HPLC.

Source: https://www.waters.com/waters/en_US/How-Does-High-Performance-Liquid-Chromatography-Work%3F/nav.htm?cid=10049055&locale=en_US

The resolution will be maintained in many conditions. Most of the reactions occurring in this technique are dependent on the recognition forces. The base of HPLC is dependent on the orientation of the proteins and peptides at the stationary phase. Most of the biological processes are based on the molecular interactions and this property can be used in the affinity chromatography. Some of the specific interactions can be used in the purification of the biomolecules and it is based on the biological functions and the chemical structures. The hydrophobic interaction chromatography or HIC, ion exchange chromatography or IEC and reverse phase HPLC or RP HPLC (Figure 4.2) can be used to separate the proteins and peptides. This is dependent on the surface charge and hydrophobicity on the surface.

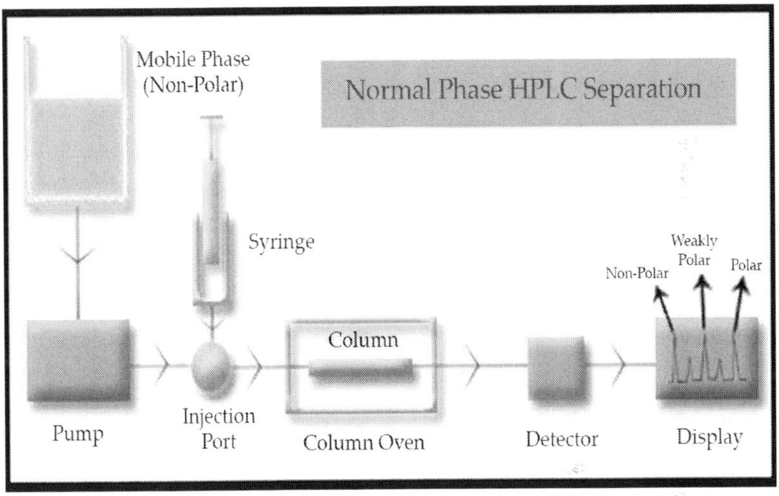

Figure 4.2. RP-HPLC.

Source: http://project-a-maize-ing.blogspot.com/2014/02/reverse-phase-high-performance-liquid.html

This technique can be used to separate complex mixtures. In the case of affinity chromatography, we can separate a smaller number of components from a sample mixture.

IEC and Size exclusion chromatography or SEC can be used to separate the peptides. Reversed -phase chromatography or RPC can be used for the same purpose. The proteins in their 3D structure can be sensitive to rough conditions which are common in RPC. This process can be used in protein isolation in a few cases. It is important to observe that the recovered proteins are biologically active. HPLC can be used in the isolation of proteins and peptides from both biological sources and synthetic sources. The applications of HPLC are - growing.

4.2 PRINCIPLE OF SEPARATION

The separation of proteins and peptides from a mixture can be due to differential adsorption of respective solutes. This is based on affinity also. The degree of binding affinity is based on the immobilized ligands and structure of the solutes. In the case of HIC and RPC, binding occurs via hydrophobic interactions. In the case of IEC, the binding affinity is based on electrostatic interactions. The time needed for a solute to pass via column can be called as the retention time. The factors that regulate the resolution of proteins and peptides help us develop different protocols to get the separation. High resolution in the separation can be based on the components in the sample and by regulating the solute peak shape and retention times. Protein and peptides can be analysed in details using several separation techniques.

4.3 REVERSED-PHASE HPLC

RP-HPLC is used to separate the molecules according to their hydrophobicity. The separation is primarily based on the hydrophobic binding of the solute to the hydrophobic ligands that are immobilized. The solute mixtures can be applied to the sorbent through aqueous buffer. The solutes did eluted with the help of solvents. The elution process can be due to isocratic conditions. In the case of the elution via isocratic conditions, the solvent concentration must be kept constant. In the case of the gradient elution process, the volume of the organic solvent should be constantly increased with time. The solutes can be eluted due to hydrophobicity. The RP-HPLC technique is a powerful tool because of several advantages. We can expect the best reproducibility of the separations. There will be high recovery rates and so the productivity will be increased. The chromatographic selectivity can be altered. We can do this by altering the mobile phase. We can get high resolution even at diverse conditions. RP-HPLC can be yield irreversible denaturation of the samples. This will reduce the biological activity of the product. There will be multiple peaks even with the pure samples, low protein yield and the enzymatic activity will be lost. All these factors will lead to the inactivation of proteins. This technique can be versatile at least for purification of proteins and peptides. This technique can be used for both preparative and analytical applications. We may use preparative RP-HPLC for purification of proteins so that we can perform sequencing. This technique can be used in the initial analysis and purification on at a large scale too. The synthetic peptides can be purified but there will be initial separation. The separation mostly will be analytical in nature. The isolation of proteins from a biological fluid or tissue extract may need a combination of techniques to yield homogeneous samples. We can use HPLC techniques at the final stages of the purification process. It is important to decide the selection of the initial separation parameters and their optimization.

4.4 MATERIALS & METHODS

The chemicals used in this technique are trifluoracetic acid or TFA, Milli-Q water and Acetonitrile or CH_3Cn-HPLC grade. The solvent delivery systems must have a binary gradient and it must be fitted with a UV detector. We need reverse -phase octadecylsilica. The solvent filtration apparatus must be fitted with a Teflon filter (0.22 micrometers). We need sample filters and C18 guard columns. Two buffers are needed during the HPLC run. The buffer A contains 0.1% of TFA dissolved in water and buffer B contains 100% acetonitrile with 0.1% TFA.

We need to add 1mg of a sample in 1 ml of buffer A solution. In the case of undissolved materials detected, we need to go for filtration via 0.22 micrometers filter. All the solvents must be filtered prior to their use. Due to this filtration, all the particulates that may block the column and solvent lines may be prevented. The filtration process can be used in the degassing of the solvent. The degassing equipment must be installed with HPLC and need to be connected online. The column and guard should be connected to the solvent delivery system. This connection must be according to the HPLC system requirements. Some of the initial conditions are ambient temperature, detection wavelengths (215 nm), flow rate (1mL per min) and 100% buffer A solvent. We need to inject 10μl of Milli-Q-water with an automatic injector or manually. It is important to use 2 blank runs so that the column equilibrium shall be maintained. We need to ensure a stable baseline. 10μL of a sample via automatic injector or manually can be injected. We need to ensure a linear gradient. The linear gradient will start from 0% to 100% buffer solution. These steps will continue for 30 minutes so that the samples will be eluted. Most of the components generally get eluted in the gradient time. Every method can be straightforward so that the rate of ambiguity will be reduced. In the analysis of proteins and peptides via RP-HPLC, we need to use a mobile phase and

C-18 based sorbent. Porous silica can be used as packing materials. It can be used to ensure greater linear flow velocities. There will be an effective mass transfer. We can perform rapid analysis. The silica can be altered by silane. The silane have hydrophobic ligand. C18 can be used as a ligand in most of the cases. We can use C8 and C4 ligands so that there will be high selectivity. The desired level of sample and efficiency decides the column dimensions. In the case of low molecular weight proteins and peptides, we need to increase the length of the column so that the resolution can be increased. The length of the column can be upto 25cm. In the case of larger proteins, there will be a loss of biological activity and mass recovery will be reduced. There will be denaturation and irreversible binding. We can use the sorter columns in this technique. The size of the column can range from 2 to 20 cm. The selection of the column diameter can be based on the detection sensitivity and sample capacity. The pore size and particle diameter can be crucial in the packing materials. When the diameter of the particle is decreased, there will be higher resolution. In the case of tryptic mapping, we need to use column lengths up to 25cm. In the case of preparative scale methods, we need to use particles that range from 10 to 20 micrometers. In the case of RP-HPLC sorbents, the pore size can be important. In the case of peptides, the pore size must be limited to 300°A. The pore size is based on the peptide size. In the case of protein separation, the pore size can be greater than 300°A. The molecular diameter of the solvent must be $1/10^{th}$ of the pore diameter. This is to prevent the diffusion of the solute. With this process, the sorbent's surface area can be accessible. This technique can be done in the mobile phase that is acidic in nature. TFA can be used as it is volatile in nature. High rates of sensitivity can be achieved by adjusting the TFA concentration in the solution of buffer B. This technique can be used so that we can manipulate the resolution and solute retention times. This can be done by certain modifications in the mobile phase. The retention of both protein and peptide retention can be due to interactions in the ligands. We can use 2-propanol, methanol and acetonitrile as organic solvents in RP-HPLC. Using these solvents can yield high rates of optical transparency. This can be used in the identification of wavelengths that are used in protein and peptide analysis. 2-propanol can be classified as strong eluent and acetonitrile can exhibit low viscosity. The composition and nature of the organic solvents can affect the proteins and peptides. The nature of the solvents can affect the selectivity and recovery rate of the materials that have biological activity. In the case of an analytical scale column, the flow rats will be upto 2 mL/min. In the case of microbore columns, we can expect greater flow rates which can be upto 250μL per min. We can detect the peptides and proteins at 210 an

220nm. We can detect tyrosine and tryptophan at 280nm as they are classified as aromatic amino acids. We can use photodiode array detectors. The detection tendencies can be enhanced by connecting these detectors online so that we can get solute spectra. We can observe the peaks in the spectra. The peaks correspond to the spectral characteristics that can be used to validate the purity. We can use spectroscopy to get information about the conformational integration of the proteins. The operating temperature plays a crucial role in determining the resolution or even manipulating resolution. The separation of proteins and peptides can be performed at normal room temperatures. The change in the temperatures can affect both the retention times and structure of the peptide mixtures. The set of gradient conditions can be based on the molecules which are to be isolated.

4.5 ION EXCHANGE CHROMATOGRAPHY

Ion exchange chromatography (Figure 4.3) is known to differentiate the biomolecules like polysaccharides, charge carbohydrates, polynucleotides, nucleic acids, polypeptides and proteins. The separation will be based on the charge. Ion Exchange Chromatography or IEX can be classified as a selective chromatographic technique.

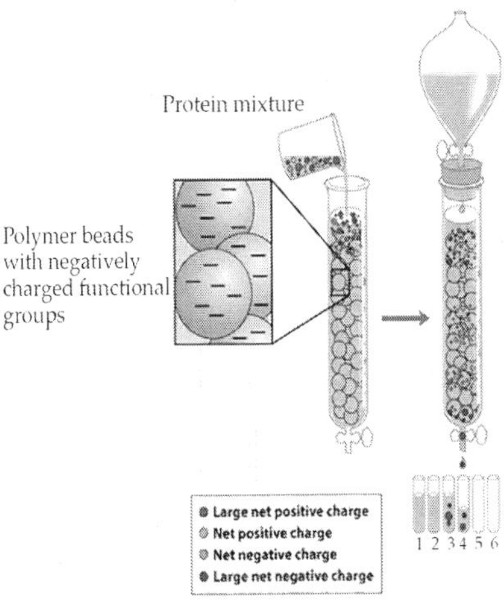

Figure 4.3. Ion exchange chromatography.

Source: https://www.intechopen.com/books/column-chromatography/ion-exchange-chromatography-and-its-applications

This process is based on ionic bond formation. The non-bonding molecules is classified as neutral molecules and they may not have charges. They are eliminated by washing. The molecules that are bound can be recovered via elution buffer. We can alter the pH of the solution in the elution process. There are some advantages of the IEX technique. This technique can be employed whilst the concentration of the proteins. The buffer components are compatible or nondenaturing with the assay systems. We can expect high recovery rates. Faster separations will be possible. We can expect a high resolving power. There are some disadvantages to IEX. We need to add an extra buffer. The instruments used in this technique should not be sensitive to corrosion. There will be high salt concentrations. There are 2 types of ion exchangers. They are an anion-exchanger and cation exchanger. In the case of anion exchanger, there will be a positive charge on the gel matrix and the anions will be in the solution. These charged groups can be immobilized on the supports. IEX can be used in batch mode and it is very simple to use.

These groups can be classified into strong and weak ion-exchangers. This classification is based on the pH range. The strong ion exchangers comprise of sulphonic acid groups that can be used in the cationic exchange. It also comprises of quaternary amino groups that are used in the anion exchange. These exchangers can be ionized over huge pH ranges. The weak ion exchangers can remain ionized at narrow pH variations. The strong ion exchanges can be pH- independent. The range of pH will be from 2 to 12. The selection of media in the ion -exchange chromatography can be based on the features of the desired biomolecules. Some of the characteristics are pH stability, molecular size and pI. At the initial stages, we may use a strong cation exchanger or strong anion exchanger. These exchangers support the loading capacity even in different pH ranges. Once the primary experiments are successful, then we can go for weak ion-exchangers. The supports used in the IEX can vary in their mechanical stability. In the case of low-pressure chromatography, we can use soft supports like cellulose, dextran and agarose. We can use these supports in the case of open column chromatography. While using these supports, there will be minimal linear flow rates. We can shrink and swell these supports. It will be based on the ionic strength and pH required by the eluent. In the case of high-performance media, we use small beads that are made from benzene derivatives, polystyrene and silica. They may not shrink or swell at high pressures. These supports can be used HPLC. Mono-S and Mono-Q may not shrink or swell and they are stable even at high pressures. The Silica- based packing should be used in the pH range between 2 to 8. The basic pH values can be used in disintegration and hydrolysis. We can observe different pore sizes in the supports used in the IEX supports. We can use supports that have less pore size. These supports can be used to elute peptides and nucleotides. The supports with large pore size can be used in the case of nucleic acids, proteins and peptides. The selection of support can be based on several factors. Some of them are size and type of particles, capacity, cost and porosity. If the size of the particles is 10 μm, then we can ensure high resolution and so there will be high backpressure.

The rate of flow will be diminished and this demands advanced pumps and hardware. In case, the size of particles is large i.e., upto 50 μm, we can expect high flow rates. There will be low back pressures. The resolution will be compromised in the case of larger particles. The strength of the supports used in ion-exchange chromatography gets increased with the size of the particles.

The strength of the IEX supports can be based on many variables. Some of the variables are flow rate, ionic strength, buffer pH, molecular size and charge. We can use the product information provided by the manufacturer. Some of the details provided by the manufacturers are dynamic capacity, available capacity and total ionic capability.

The number of charge groups available on the support can be called as total ionic capacity. All the ion exchangers can be porous in nature. The shape and size of the proteins can determine the accessibility of them towards charged groups. The amount of a protein that can bind to anion-exchanger at certain sample concentration, ionic strength and pH can be called as available capacity.

The available capacity can be more than the dynamic capacity at certain conditions. There will be binding of sample to the support whilst the buffering ions are not competing.

Most of the buffering ions must have the same charge as the ion exchanger. We can use cationic buffers for anion exchange and anion buffer for cation exchange. The buffer additives like protease inhibitors and detergents may not be charged. They may sometimes have the same charges as the ion exchanger.

The chloride ions can be employed as a counter ion in the case of anion exchangers.

We can use sodium as a counterion. It is important to use distilled water and deionized water in the preparation of buffer. The buffer should be degassed and filtered. These buffers can be used in the HPLC columns. The pH of the buffer must be maintained accurately. The pH electrodes must be calibrated frequently. The calibration can be performed with pH standards. We need to monitor the volume of both acid and base. The IEX based separation can be performed using batch technique. The exchangers can be packed in the column.

We can use batch processing to select and separate the proteins into two groups. They are the bound group and nonbound group. The elution of the bound components can be achieved by the support and it can be done by enhancing the salt concentrations. The buffer pH can be changed in this case. The concentration of salt can be increased gradually. The elution process in the linear gradients needs an appropriate gradient controller and pumping systems. Greater resolution can be expected from column chromatography. It is important to know that IEX can be the best choice in the purification process. The linear gradient can be used to determine the steepness of the gradient. It can be considered as a measure to estimate the rate of the counter ions whilst separation.

The steepness of the gradient can be determined by the volume of the column, gradient time, salt concentrations and rate of flow. Most of the shallow gradients can yield an optimal resolution. We can get gradient yields at faster rates and we can expect reduced band broadening. The concentration of the salt required to elute the substances from the ion exchange column can be based on the exchanger type and this is true whilst dealing with the unbound substances.

4.6 HIGH PERFORMANCE HYDROPHOBIC INTERACTION CHROMATOGRAPHY

Hydrophobic interaction chromatography or HIC (Figure 4.4) can be classified as a column chromatographic technique that is also used in protein purification. We can separate the polynucleotides and proteins. Most of the purification systems can be enhanced by adding HIC into their systems so that the productivity will be increased. We can use HIC for affinity chromatography, size exclusion chromatography and ion -exchange chromatography. HIC can be used in the purification of the proteins which can be present on the membranes, recombinant proteins, cells, receptors, polynucleotides, nuclear proteins and serum proteins.

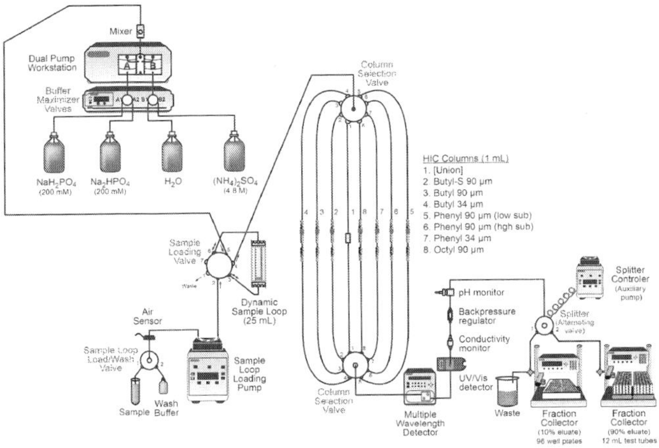

Figure 4.4. Hydrophobic interaction chromatography.

Source:https://journals.plos.org/plosone/article?id=10.1371/journal.pone.0108611

HIC can be combined with the nondenaturing properties of salt in the chromatographic process. With this technique, we can yield high resolution and the recovery will be good. This technique is based on the adsorption principle. The biomolecules will be adsorbed on the hydrophobic surface whilst high salt concentrations. We can perform the elution process via a salt gradient. This technique was first used in the separation of proteins and it was done by soft gel media. Due to the introduction of HPLC, we can use silica. The conditions pertaining to HIC can be the same as the salting protein purification methods that are conventional. The chemicals used in this technique are ammonium sulfate, Milli-Q water and 0.1 potassium phosphate buffer at neutral pH.

We may need HPLC solvent delivery system with a UV detector and it is essential to maintain binary gradient capability. The hydrophobic interaction column can allow 5 µm particle size. Most of the supports used in IEX are prepacked columns and preswollen gels. Whilst using dry powder supports, we need to allow it to swell. The preswollen gels can be used in ready-made form. It is important to mix the gel with slurry. We need to take some portion and add it to the vacuum flask. We need to allow it to settle. Post chromatographic run, we need to wash the column and it is important to store it in the storage solvent. There are some limitations to this technique. The column may not be completely equilibrated with the buffer. The sample may not be present at accurate pH and in some cases, salt may be present.

The starting buffer may have too much salt concentration. The starting buffer may not have accurate pH. The rate of solubility of several proteins can be low at their isoelectric points. It is important to maintain appropriate pH conditions.

Due to this, there will be isoelectric precipitation in the columns and this will block the columns. The recovery percentage will be reduced. The level of two buffers must be the same and the mixture must be set to ON in all conditions. In case there is continuous mixing, the buffer levels in the chamber will rise and there will be backflow. It is important to use the gradient marker. There will be overhead stirrer. It is important not to use magnetic stirrer. The magnetic stirrer cleave the gel particles. It is important to avoid the direct mode of application. At alkaline pH, a minimal sample will be bound in the case of a pilot test. The IEC can be classified as an adsorption technique. In this technique, we can load large volumes of samples onto the column. The sample can be eluted via elution buffer.

4.7 GEL FILTRATION CHROMATOGRAPHY

The gel filtration chromatography (Figure 4.5) can be called size exclusion chromatography. It can be used to separate the molecules based on their molecular weights or sizes. In this technique, we can use gel media and it is suspended in the buffer solution. This gel can be packed onto the column. These columns can be of different sizes. The gel media can comprise particles that are porous in nature. The small molecules can be diffused freely via pores.

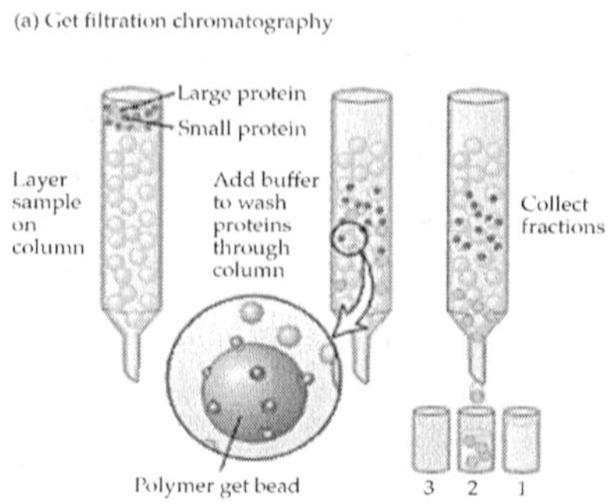

Figure 4.5. Gel filtration chromatography.

Source: https://www.slideserve.com/alton/gel-filtration-chromatography

The movement of molecules will be lowered in the column. There will be a separation of the molecules and it will be based on the reducing molecular weight. The molecules withlarge size will be eluted first. The gel filtration chromatography can be classified into two classes. They are group separation and fractionation. In the case of group separation, there will be a separation of dense molecules from less dense molecules. In the case of fractionation, the molecules will be separated as per the size variations and this is important for purification protocol.

The selection of media in this technique is based on the applications. It is important to know the molecular weight of the solute. The tendency of the gel filtration column can separate two components that can be proportional to the square root of the length of the column. The factors needed to determine the dimensions of the columns can be based on the sample volume. In normal conditions, the eluent composition may not be used to determine the separation. The samples yielded from this technique can be concentrated before processing. The concentration process can be performed by freeze- drying. The buffers that are volatile in nature can be considered as an alternative. The acidic conditions can help us to dissociate protein complexes. Most of the buffers used in this technique must be prepared afresh and it is important to use degassed water. Due to this process, the bubble formation will be prevented. The bubbles may form in the columns and tubing lines. Most of the buffers should be degassed if they are to be used after 48 hours of storage. The flow rates can be analyzed using the manufacturer's manual. The flow rate will be higher if we use specific gels and it is based on the size of the particles. We can get enhanced resolution only if we get minimal flow rates. The proteins can be detected using UV radiation (280nm). Most of the UV detectors will be installed below the column. There will be a fraction collector in the downstream processing. The gel can be in the form of dry powder. In this case, we need to measure the gel weight. It is important to determine the bed volume of the columns.

4.8 HPLC

HPLC can be used to separate, identify and quantitatively measure the components present in the mixture. The liquid solvent with the sample will be pressurized to flow via a column. The column will be an adsorbent material. Every component present in the sample can be linked with the adsorbent material. There will be varying flow rates with different compounds. There will be component separation whilst they flow.

The chromatography can be classified as a mass transfer phenomenon and it is nothing but adsorption. In HPLC, we need to use a pump so that the sample and the solvent liquids will be passed through the adsorbent with varying associations. The column is active and it has adsorbent. The adsorbent can be made up of silica or any other polymers. It has granular materials. The size of these particles will be upto 50 μm. There will be different degrees of association with these particles. Because of this, we can separate various components of the sample. The pressurized liquid will be a solvent mixture. The solvent can be methanol, acetonitrile or water. This can be called mobile phase. The temperature and composition play a vital role in the entire separation process. There will be various interactions between the adsorbent and the components of the sample. Most of the interactions can be physical. The interactions can be ionic, dipole-dipole and hydrophobic in nature. The pressures in the operation of HPLC can be upto 350 bar. In conventional chromatography, we can expect the pressure only due to gravity. The average column dimensions of HPLC are 30-250 mm (length) and 2.1 to 4.6 mm (diameter). Most of the columns used in HPLC are produced by minute adsorbent particles. We can have an average particle size of 2 to 50 μm. Due to this feature, we can expect high resolving power. Due to its high resolving power, we can classify HPLC as a powerful chromatographic technique.

In HPLC, there will be sampler, pumps, degasser and detector. We can add a sample mixture onto the mobile phase. Through the mobile phase, the sample will be supplied to the column. The pump can be used to supply the required pressure. The pump ensures that there is correct flow and the mobile phase is accurately passing via column. The detector can produce a signal. This signal can be equal to the sample amount which was eluted through the column. We can use this technique in quantitative analysis. We can use a digital microprocessor for acquiring data. In some HPLC models, we can use multiple solvents. These solvents can produce gradient, especially in the mobile phase. Some of the detectors used are mass spectrometry, photodiode array, etc.

The sample volume is injected into a mobile phase that can migrate via column. The sample pass via column with varying velocities. These velocities explains the physical interactions between the sample components and adsorbent. Each component exhibit a specific chemical property. The velocity can be based on the mobile phase composition, stationary phase and chemical nature of the sample. The time within which the analytes elute can be called as the retention time. We can identify the analytes with the help of retention times. There are several column types. Most of these columns will be filled with adsorbents. These adsorbents will be of varying particle sizes. The nature of the adsorbent surface plays a vital role. The small size of the packaging materials may require high backpressures. Due to this, there will be an enhanced resolution. The sorbent particles can be polar in nature and some particles can be hydrophobic in nature. The mobile phase constitute water and organic solvents. We can use methanol and acetonitrile as solvent. The HPLC techniques can be sometimes water-free. Most of the aqueous compounds can comprise of acids. The acids that can be frequently used are trifluoracetic acid, phosphoric acid and formic acid. We can use salts in the component separation. In some cases, the mobile phase composition can be constant. We can call this a isocratic elution mode. This mode can be called gradient elution mode. This mode is effective in component separation. In the case of gradient elution, the mobile phase composition differ from the eluting strengths. This can be affected by retention times. The mobile phase composition can be called as eluent. The elute composition is based on the interactions between the stationary phase and analytes. The analytes are nothing but the sample components. The selection of gradient conditions, additives and mobile phase components depends on the sample components and column.

Before HPLC, researchers used liquid chromatographic techniques and they have many limitations. Most of the solvent flow rates are based on gravity. In this technique, the separations were slower. Gas chromatography or GC was considered as a better technique than liquid chromatography or LC (Karger et al., 1997). The thermal stability of the solutes if lower in the case of GC (Henry et al., 1965). Most of the history of HPLC was based the particle technology. Some of the problems with early- stage HPLC are that there was harder to produce uniform packing and the pressure drop required to move the fluid via column (Giddings et al., 1965). Partition chromatography was considered as an efficient technique then (Ettre, 2001). The principle used in the case of paper chromatography was the partition coefficient principle. The same principle was applied to liquid-liquid separations, gas phase applications, thin layer chromatography, etc. Normal phase HPLC or NP-HPLC was used to separate the analytes. This separation was dependent on their affinity towards silica. In the case of displacement chromatography, one molecule with a higher affinity for a displacer can compete with others for binding sites. The molecules with high affinity can displace the molecules with lower affinity. The reverse phase HPLC or Rp-HPLC can have a stationary phase that is non-polar in nature and an aqueous mobile phase. Silica can be used as a stationary phase. The retention time can be longer in the case of less polar molecules. Most of the polar molecules can elute at a faster rate. The RP-HPLC work due to the hydrophobic interactions. The size exclusion chromatography can be called gel permeation chromatography and it separates solute components based on the molecular sizes. To be exact, we mostly use the Stokes radius. This technique may yield a low resolution.

4.9 OPTIMIZATION STRATEGIES

The optimization of the processes is important in any chromatography technique. The speed at which the HPLC runs can be considered as a basic criterion. The separation of solute components in the least possible time decides the optimization process. If the HPLC unit elutes the sample components at high speeds as defined, we can say that the technique and process being optimized. The increase in the speed of this technique determine the economics of the overall technique. The speed of this technique also determine productivity. The resolution between the peaks can be considered as an optimization goal. We can consider two approaches that affect chromatography. They are 1. Improving the selectivity and 2. Increase in the column efficiency. The resolution must be optimized so that the precision and accuracy can be enhanced in the case of quantitative analysis. Due to this, the robustness will be increased. Enhanced separation efficiency can translate into an increase in components. We can achieve theoretical peak capacity with this technique.

The limit of detection or LOD can be defined as the minute component of the analyte that can be observed. The LOD value can be classified as a vital parameter in the case of trace analysis. The plate number can be considered as an important measure to determine the separation efficiency of the column. This can be measured under certain temperature conditions and the mobile phase is used. Generally, the plate number will be proportional to the length of the column. In case there is an increase in the plate number, we can easily separate analytes with limited effort. The retention factor is also called retention parameter. It is not the same as the retention time. It is based on the temperature, mobile phase conditions and stationary phase conditions. We can obtain a retention factor from peak retention times. The separation of the compounds depends on relative retention. This can be obtained from two peaks and we call it a separation factor.

4.10 BIOCHROMATOGRAPHY

Biochromatography (Figure 4.6) can be used in the pharmaceutical industries in the analysis, development and production of ingredients. They are called as active pharmaceutical ingredients or APIs. Due to this technique, there will be an increase in biomolecules. Biochromatography can be used in case of smaller molecules with molecular weights less than 1000 Da. This technique uses polymer-based materials. In this technique, we use diverse stationary phases. The base material can be chemically stable against solvents, bases and acids. The stationary phase can be determined by several parameters. They are functional groups, purity, bonding density, surface area, pore size, size and form of particles.

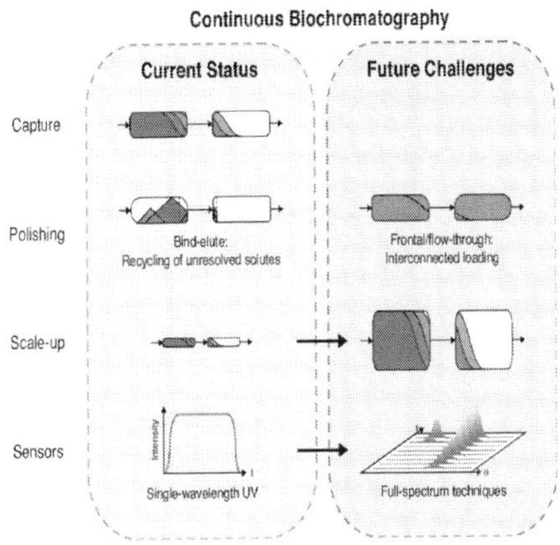

Figure 4.6. Biochromatography.

Source: https://ars.els-cdn.com/content/image/1-s2.0-S2211339818300327-fx1_lrg.jpg

We can consider three -particle forms like spheroids, spherical and irregular forms. The irregular particles can be produced via the milling of massive particles. The spherical materials can be packed easily. The spheroidal forms are rare. The size of the particle is nothing but the mean diameter of the particles. The size of the particle determines the pressure and plate number. The increase in the plate number is possible by using small particles. We can also increase pressure with an increase in the plate number. In this technique, the size distribution of the particles can be important. An increase in small particles can enhance the backpressures. In the case of large particles, the plate number will be decreased.

The use of soft gels and small particles do affect the field of application i.e. the length of the column and flow rate. While analyzing the biomolecules, it is important to use an accurate pore size. The use of small pores can be beneficial as they offer high surface area. There will be high loading. We can define the distribution and size of the pores through nitrogen adsorption or BET. We can also use mercury porosity for that matter. The loading density can be considered important in choosing the stationary phase. The dynamic binding capacity can be based on the size and type of sample components. Pore accessibility and flow rates can also determine the dynamic binding capacity. The biomolecules can be classified as a multifunctional unit. There will be interaction between them. There will be some impurities that can interact readily with the samples. The expected purity must be at least 99%. The metals like Fe and Al can be present in the silica gels. Due to the polymerization of the products, there will be impurities for certain. Most of the functional groups are based on the type of chromatographic technique being used. The reversed- phase chromatography can be classified as a method of choice to separate most of the proteins and peptides. Acetonitrile can be considered as the best organic modifier in the case of preparative and analytic separations. TFA can be used as an ion pair reagent. The particles must be small and it will offer narrow peaks and short diffusion paths. There will be enhanced backpressure. The size distribution of the pore can be important in the separation of the proteins and peptides. The size distribution of the pores can be broader. Due to this, there will be a separation of components based on the selectivity. We can compensate for this by using longer columns. The volume of the pores can determine the elution volume and run time of separation. In the case of higher plate numbers, there will be reduced flow rates. Due to high temperatures, there will be lowered viscosity and so we can expect a high rate of diffusion. Reduced viscosity can enhance efficiency. Most of the viscosity can be dependent on organic modifier and salt concentrations.

The human insulin and porcine do vary by one amino acid. Insulin from rabbits can be resolved from human insulin by only one amino acid (Rivier, 1983). The separation process could be due to the adsorption and desorption process. It is important to use an organic modifier so that there will be desorption of the proteins (Christianson, 1994).

Small pores can be about 100 angstroms in the case of silica. It is used in the case of RP-HPLC. In the case of coarse protein separations, the pore silica used will be of 300 angstroms (Pearson, 1982). Silica can be altered by a hydrocarbon molecule. This alteration can yield a hydrophobic surface.

The modified silica surface can attract organic groups to the silica. The columns can be used in separating the polypeptides (Tempst, 1986). The selectivity of peptides can be most affected by the nature of the phase and the silica surface. There are different types of silica being manufactured and each type of silica offer different selectivity. The diameter of the analytical, narrow bore, microbore and capillary columns can be 4.6mm, 2mm, 1mm and less than 1mm, respectfully (Davis, 1992). The capillary type of column can exhibit higher relative sensitivity of 200. Analytical columns exhibit a relative sensitivity of 1 (Davis, 1992). Ethanol and methanol can be used as organic modifiers (Sussman, 1988). The resolution of peptides can get reduced due to change in the solvent concentrations and this yields coelution (Chloupek, 1992). We can use heptafluorobutyric acid to resolve basic proteins like histones (McCroskey, 1987). We can use mass spectrometry post chromatography on-line. The mass spectrometry can measure the mass and charge. We can identify the fragmentation data to detect the peptides. This technique can identify peptides and proteins based on mass. It also gives us information about the separated proteins and peptides along with their molecular weights. It can be classified as a perfect detection technique (Day et al., 1997).

The electrosprays can give us a high rate of shear flow and they are called pneumatic assisted electrospray. They can increase the flow region (Hopfgartner et al., 1993). We can use formic acid in place of TFA (Kumaresan et al., 1995). The enzyme protease can be useful in both quantification and identification of the proteins. We can use trypsin as a protease and we need to use 5 steps in the protein digestion process mediated by trypsin (Lundell et al., 1999). The mapping of peptides can be linked with mass spectrometry and we can determine the location of the disulfide bonds (Mhatre et al., 1999). We can use columns that are 15cm in length and 60 cm in diameter and they can be used in the purification of the proteins (Rivier et al., 1984). The recombinant erythropoietin can be purified using RP-HPLC (Por-Hsiung, U.S. Patent 04667016). Recombinant insulin or humulin can be purified via RP-HPLC in its active form (Kroeff et al., 2020).

4.11 APPLICATIONS

Abderhalden and Muller have described the difference in the activity of 2 enantiomers of a drug (Abderhalden et al., 1908). They showed that the (R)-Adrenaline and (S)-Adrenaline act differently on the blood pressure. In current situations, it is known that enantiomers can differ in their pharmacological and pharmacokinetic properties. The enantiomers differ in their excretion behavior, metabolic behavior, receptor binding, distribution and bioavailability (Caldwell, 1996). The popular techniques that are used in chiral separations are Thin Layer Chromatography (Figure 4.7), Supercritical Fluid Chromatography (Figure 4.8), Gas chromatography (Figure 4.9), HPLC, Capillary electrophoresis (Figure 4.10) and capillary electrochromatography.

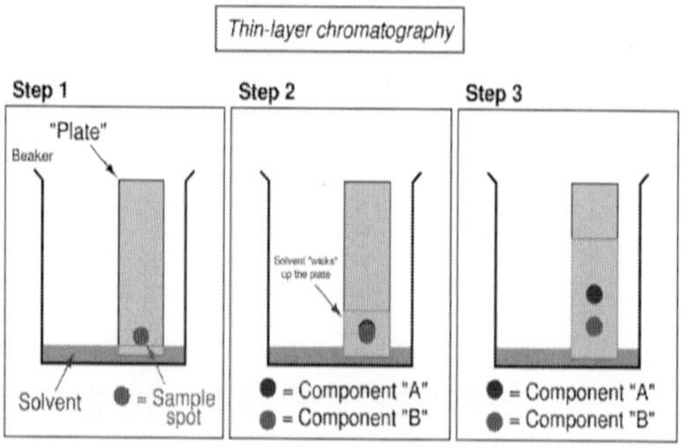

Figure 4.7. Thin Layer Chromatography.

Source: https://instrumentationtools.com/thin-layer-chromatography-manual-method/

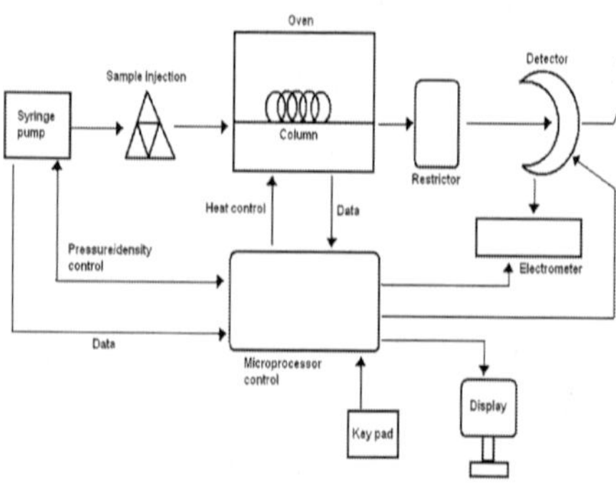

Figure 4.8. Supercritical Fluid Chromatography.

Source:https://chem.libretexts.org/Bookshelves/Analytical_Chemistry/Book%3A_Physical_Methods_in_Chemistry_and_Nano_Science_(Barron)/03%3A_Principles_of_Gas_Chromatography/3.03%3A_Basic_Principles_of_Supercritical_Fluid_Chromatography_and_Supercrtical_Fluid_Extraction

The chiral separation methods that are linked with HPLC are classified into 2 classes. The chiral separation classes can be direct and indirect. The direct method is dependent on the formation of a complex. The indirect method is dependent on covalent bond reactions (Srinivas, 1992). The chiral stationary phase can be comprised of polymers and minute chiral molecules (Yashima, 2012). Huang et and his coworkers developed an advanced method to determine the purity of the enantiomers (Huang, 2006). HPLC was used in the chiral drug analysis via bioanalytical methods linked with fluorescence and UV. We need to avoid explosive or flammable solvents (Bakhtiar et al., 2006). The miniatured systems can speed up the process. Atenolol in urine was analyzed via Nano-LC with UV and MS (Orazio et al., 2016).

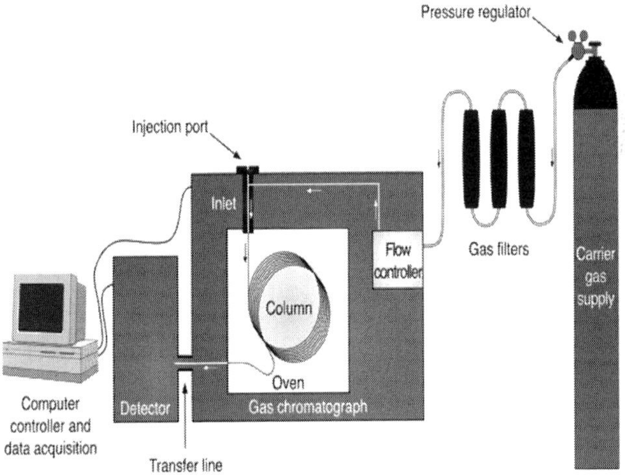

Figure 4.9. Gas chromatography.

Source: https://www.sciencedirect.com/topics/agricultural-and-biological-sciences/gas-chromatography

Most of the accurate environmental analytical methods are based on HPLC. The biogeochemical cycles are modified by the anthropogenic species. Most of the interactions with the biogeochemical cycles can be complex (Larson et al, 1994). The metal speciation in the soil and surface water can be analyzed using HPLC-MS methods (Terlecka, 2005). The pesticides can suffer from the degradation processes (Medana et al.,2005). Aromatic amines are present in the environment due to man-made activities.

The HPLC technique can be used to concentrate the aromatic amines from the polluted waters (Rao et al., 1999). In the case of mass spectrometry, the amines can be detected even with low voltages (Ballantine et al., 1997). RP-HPLC can be used in monitoring the wastewaters (Rehorek et al, 2002). We can use RP-HPLC to measure the carbonyl compounds, ketones and aldehydes (Muller, 1997). The nitrophenols that are toxic can be analyzed using HPLC (240). Most of the nitrophenols in the snow and rain can be quantified using HPLC linked with UV (Schussler et al., 1998).

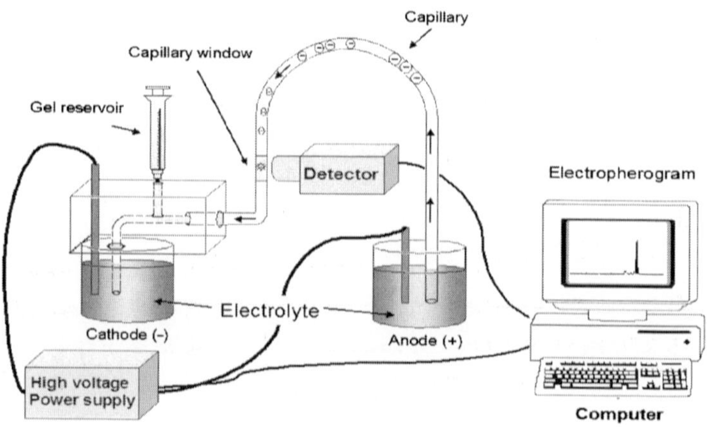

Figure 4.10. Capillary electrophoresis.

Source: https://link.springer.com/protocol/10.1007/978-1-60327-375-6_6

5

Electrophoresis

CONTENTS

5.1 INTRODUCTION

Electrophoresis (Figure 5.1) can be run through agarose and polyacrylamide gels and it is classified as a standard method in molecular biology. This technique can be used in separation, identification and purification of nucleic acids and proteins. The separation is based on isoelectric point, form and size. The charged molecules can migrate via gel matrix due to an electric field. This technique is important in proteomics, analytical chemistry, molecular biology and biochemistry. Mostly, gel electrophoresis can be employed for analytical needs. We can use this technique as one of the preparative techniques. We can use this technique for purifying the molecules. Mass spectroscopy can be used for further analysis of the proteins previously resolved by gel electrophoresis (Wasinger et al., 1995). We can classify gel electrophoresis as a conventional technique.

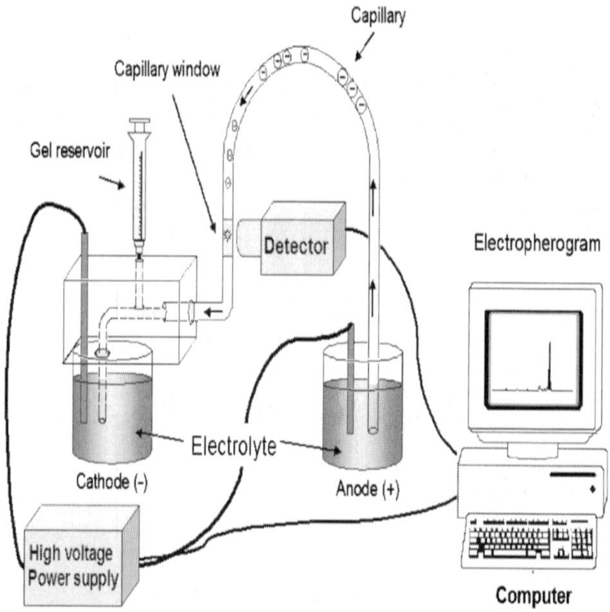

Figure 5.1. Gel electrophoresis.

Source: https://www.sciencedirect.com/topics/neuroscience/gel-electrophoresis

The gel electrophoresis can be used in resolving proteins. We use a polyacrylamide matrix in case of protein purification. This technique is called as Polyacrylamide gel electrophoresis or PAGE. This technique can be used in protein characterization. This technique can be inexpensive and convenient. Very less quantity of protein are needed in the case of PAGE. The proteins with high charge can exhibit fast migration. Most of the proteins do not exhibit predictable structure whereas the nucleic acids can exhibit a predictable structure. The proteins may not get migrated on the application of electromotive force. We can add detergent in the protein denaturation process. SDS or Sodium dodecyl sulfate can be used to separate the proteins as per their molecular weights (Shapiro et al., 1967). SDS molecules cleave the disulfide bonds and so the proteins can be cleaved into sub-units. Due to the application of SDS, there will be a negative charge all over the molecules. This charge can be used in the migration through the gel and the separation is based on the molecular weight. The denaturation process can disrupt the tertiary structure of the proteins. The velocity of the separation or migration can be equal to the size of the molecules. It is obvious that the separation is not based on the tertiary structure.

There are many advantages of PAGE. The gels can lower the thermal convection process. The heat can be generated due to the electric field. The gel can serve as a sieving medium. This will lower the migration of the molecules. These gels can reveal separations. We can easily apply stain after an electrophoretic run. These gels can be produced by polymerization of acrylamide. We need to ass cross-linking agent. Bis-acrylamide can work as a cross-linking agent. It is important to add a catalyst and an initiator. The initiator in the solidification of gel can be ammonium persulfate or APS. This chemical can produce free radicals. TEMED can mediate the polymerization reaction. The polymerization reaction can be done only by stabilizing the free radicals. The persulfate molecules can interrupt the electrophoretic run.

The acrylamide solutions must be degassed prior to the electrophoretic run.. The presence of oxygen can inhibit the polymerization process. Due to the production of heat, there will be bubble formation in the gel. The polymerization rate can be measured by the initiator and catalyst. The acrylamide/bisacrylamide and concentration of the compounds can influence the rigidity and pore size in the gel matrix. These factors can affect the protein sizes that are to be resolved. The pore size in the gel can be linked with the proportion of the acrylamide. If we use 12% of polyacrylamide, there will be small pores and in the case of 7% polyacrylamide gel, there will be large pores. We can use more buffers in case of acrylamide gel electrophoresis and this is called a continuous phosphate buffer system (Osborn, 1968). In the case of discontinuous system, the first buffer aid in protein migration.

The stacking gel comprises of large pores in acidic pH. The second gel may have small pores. These systems can be useful in resolving samples with high resolution (Westermeier, 2005). If the peptides are less than 14kDa, then they may not be resolved by conventional systems that are based on tris-glycine.

The proteins that are resolved on the polyacrylamide gels can be detected by several methods. They are dye staining and silver staining and more. We can detect proteins upto 0.6μg via Coomassie blue staining. Silver staining can be considered as an alternative staining procedure. This staining procedure can provide a higher sensitivity than the Coomassie blue staining. We can detect those proteins that are tagged with radioactive materials through autoradiography. (Figure 5.2)

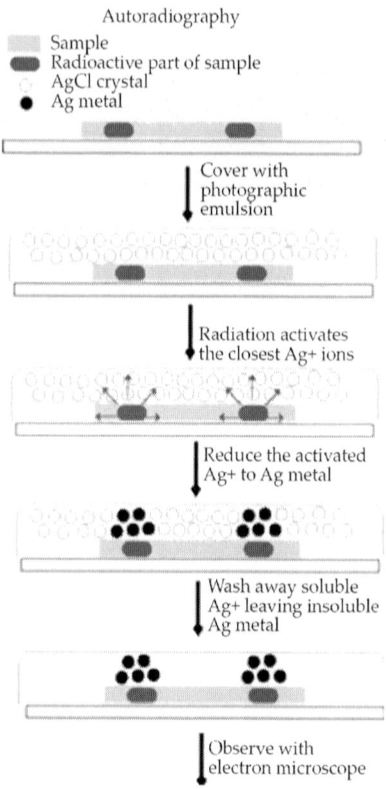

Figure 5.2. Autoradiography.

Source: http://lifeofplant.blogspot.com/2011/12/autoradiography.html

In the autoradiography, the light will be generated by the intermediate molecules. It also uses photographic emulsions. The silver emulsion can be sensitive to specific radiation. In some cases, the silver molecules can be sensitive to electromagnetic radiation and in this case gamma radiation. The metallic silver will be precipitated. Due to precipitation, there will be dark particles that are visible in the region. We can observe these darker particles at the places where the radioactive proteins are found. Based on the protein state, these techniques can be divided into native electrophoresis and denaturing electrophoresis. In the case of denaturing electrophoresis, most of the protein samples will be migrated and so there will be denaturation. We can say that the protein has lost the three-dimensional structure. The migration can be based on the size and load of the molecule. In the case of native electrophoresis, the proteins can get migrated but there will be no denaturation. In this technique, proteins can migrate as per their size, shape and charge. In this type of electrophoresis, we can use tris-acetate, tris-borate and tris-glycine buffer systems. In the case of blue native electrophoresis, we can resolve multiprotein complexes with high resolution. These proteins will get resolved in native conditions. In this technique, we can use polyacrylamide gel electrophoresis. In this technique, many samples can be analyzed simultaneously.

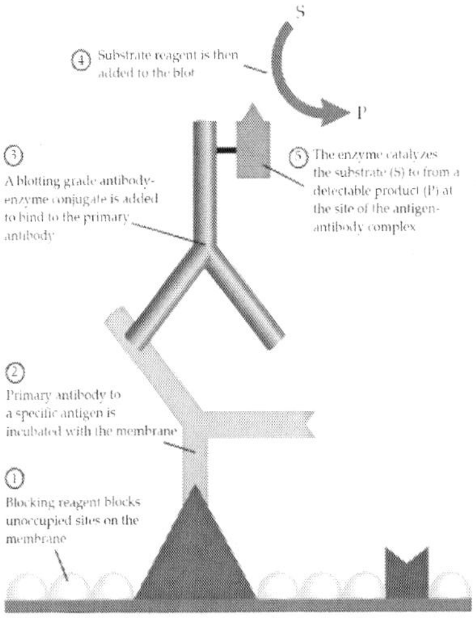

Figure 5.3. Immunodetection.

Source: https://www.bio-rad.com/en-in/applications-technologies/introduction-immunodetection?ID=LUSQ4HE8Z

The immunodetection (Figure 5.3) of the protein samples can be done with the western blot technique. This technique is being used in both biochemistry and molecular biology. We can detect proteins from the cell extract or cell homogenate. In this technique, the proteins are migrated via gel onto the nitrocellulose membrane. The membranes may be made up of polyvinylidene difluoride or PVDF or Nylon. At this stage, the proteins can be analyzed using antibodies. DNA detection can be done by Southern blotting (Figure 5.4) and RNA detection can be done by Northern blotting. (Figure 5.5)

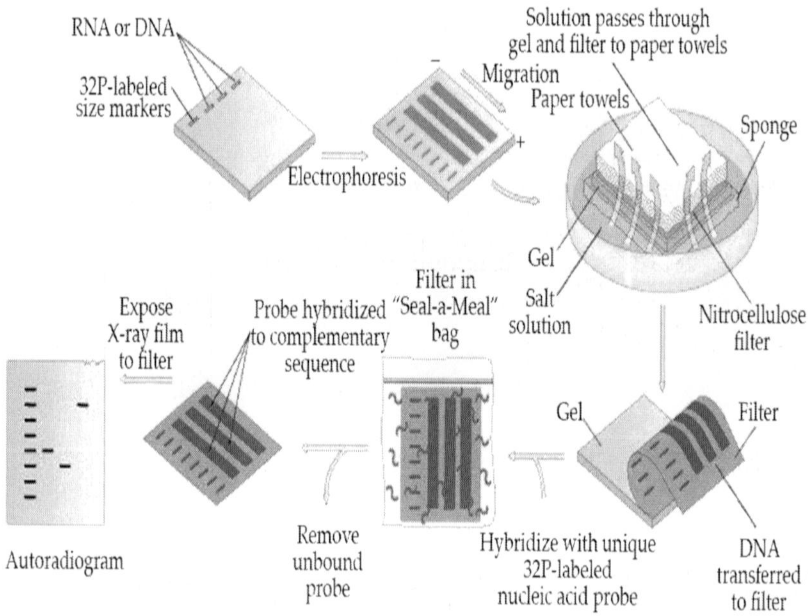

Figure 5.4. Southern Blotting.

Source: https://www.mun.ca/biology/scarr/Gr12-18.html

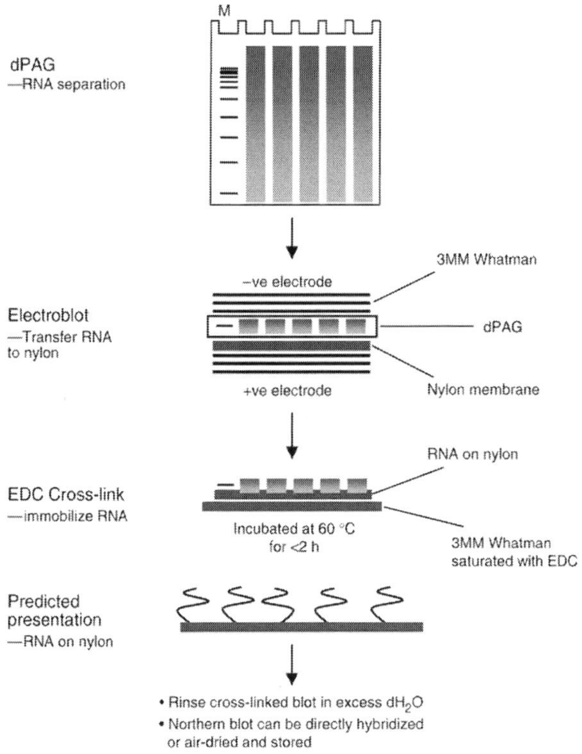

Figure 5.5. Northern Blotting.

Source: https://www.nature.com/articles/nprot.2008.67

We can transfer the proteins that were migrated via gel onto the membrane by an electric field. The electric field must be perpendicular to the gel. We can also do this process by adding a concentration solution as a dot on the membrane. Due to absorption, the proteins can stick to the membrane. So, there will be a dot or spot. This can be called as 'slot blot' or 'dot blot'. In this case, we need to apply suction so that there will be penetration of solutions. The proteins that are bound to the membranes can have several benefits. The staining process will be stronger. In the case of isoelectric focusing gels, we may not expect staining for ampholytes. We can easily detect minimal protein concentrations. Most of the membranes can be manipulated easily.

The blotting technique can be done through 5 stages. The proteins can be immobilized on the membranes. This can be done via a direct application or by transference through suction and electrophoresis. In this process, a flat sponge is piled on the filter paper and it is previously soaked in the buffer. This is piled on the gel too. So, the membrane can be direct contact with the filter paper and gel. There will be a flat sponge. This set is impregnated between 2 layers of plastic that are perforated. This is placed in a tank. In this tank, the saline solution can be considered as a transference buffer. There will be 2 plate electrodes. Due to this, there will be a uniform field on the gel. The gel will be directed toward the anode. The membrane will be toward the cathode. There will be a saturation of all binding sites present on the protein. We need to incubate the primary antibodies with the desired proteins. Incubation can also be done with secondary antibodies. Most of the proteins that are labeled with the enzymes can be visually inspected with substrates. There will be the formation of insoluble and colored products. An antibody can be conjugated with another antibody with alkaline phosphatase or peroxidase (Blake et al., 1984). We can use an amplifying enzyme. This enzyme is the biotin-avidin-peroxidase complex (Hsu et al., 1981). We can use alkaline phosphatase for this purpose. Enhanced chemiluminescence or ECL can be classified as a common technique in the detection of proteins via western blots (Laing, 1986). In this technique, there will be a production of light. The light can be emitted due to the oxidation of luminol. This reaction can be catalyzed by hydrogen peroxide / horse radish peroxides. Once the light is emitted, it will be captured by a CCD camera. This can be used in both semi-qualitative analysis and qualitative analysis. This method can include re-probing and stripping the blot with diverse antibodies.

Isoelectric focusing (Figure 5.6) or IEF can be dependent on the migration of the molecules according to the pH gradient. We can separate amphoteric molecules like proteins and amino acids. This can occur due to the differences in pH gradient and potential. The anode region can be classified as acidic and the cathode region can be called alkaline in nature.

In the case of a long pH gradient, there will be a separation of the molecules based on their isoelectric points. If a substance is below its isoelectric point, we can expect it as having a positive charge. So, this substance can migrate to the cathode. The substances that are present in media having less pI can have a negative charge and so they can move to the anode. The migration process will occur until such a point where pI coincides with pH. At this point, there will be a zero charge. There may not be any sort of migration in this case due to the formation of zwitterions. We can say that the amphoteric molecules can be present in the narrow bands. These molecules can be located where the pH coincides with the pI. The point of application may not be considered as important because the molecules can migrate to the pI regions.

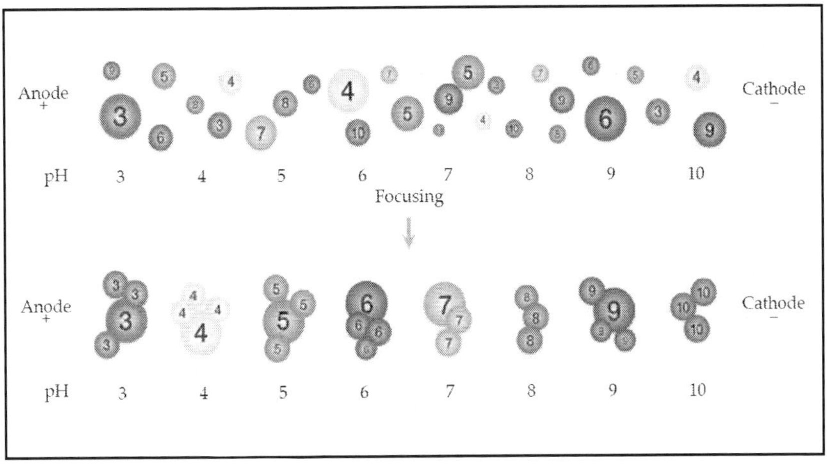

Figure 5.6. Isoelectric focusing.

Source: https://www.bio-rad.com/en-in/applications-technologies/isoelectric-focusing-2d-electrophoresis?ID=LUSQLK2B7

In the case of 2-DE or two-dimensional gel electrophoresis, there will be a separation of the protein mixtures and it is based on the molecular properties. They are properties are isoelectric focusing and molecular weights (O'Farrel, 1975). Isoelectric focusing will be between 3 to 11 pH values. The SDS PAGE has 12% of acrylamide. The preparation of samples can be based on specific methods that are important in the success of the electrophoretic runs. Some of the important factors that determine the success of the experiment are size, charge, solubility and isoelectric points of the proteins. It is important to be cautious whilst sample preparation as it will reduce the complexities. The fraction of the proteins that are to be loaded onto the gel should have a lower ionic strength buffer. This will help in maintaining the charges on the proteins. These buffers can keep the proteins soluble. Most of the proteins can be separated according to the pI and pH. It is important to include an equilibration step. In this step, the disulfide bonds can be reduced. There will be alkylation of the sulfhydryl groups that ate present in the cysteine residues. Most of the proteins are linked with SDS so that there will be a separation according to the molecular weights. We can use SDS-PAGE to resolve proteins based on their molecular weights. The gel to be used can be based on the molecular weight of the proteins. We can run many gels simultaneously at the same conditions. It can be used to compare several gels at the same. The staining process can be used to visualize proteins that are present in the gels. The choice of stain is based on different factors. They are linear range, imaging equipment type, expense, ease of use and sensitivity range. We can say that there is no universal stain available. We can perform image analysis in 2-DE. We can acquire data in the digital forms and this factor can be important in 2-DE gels. We can compare gels easily as the data is readily available. Many devices use a specialized software to collect and interpret the data from the gels. In these systems, there will be an automatic comparison of the spots so that there will be accurate and precise identification. This can be used in the quantitative analysis of proteins. The proteins after the electrophoretic runs can be excised from the gel. They can be destained and then digested. After digestion, we can identify the proteins via mass spectrometry. We call this process as peptide mass fingerprinting. Once this is done, we can compare the peptide with other data obtained online.

The two-dimensional difference gel electrophoresis or 2-D DIGE can be used to label the samples before the 2-dimensional electrophoresis. With this method, we can accurately analyze the differences present in the protein samples (Mackintosh et al., 2003). We can separate three samples that are different by using 2 DE gels. In this technology, we can use fluorescent cyanine dyes (Unlu et al., 1997). The similar proteins can be labeled with three different dyes named Cy2, 3 and 5. These proteins will move to the same position. We can use mass spectroscopy to yield high accuracy in determining the composition of diverse chemicals. Most of the materials that are to be analyzed can be ionized and they will be shifted by electric and magnetic fields to the analyzer. The ionization techniques can be used to determine the sample types that can be identified using mass spectroscopy. There are 2 important techniques that can be used in both solid biological samples and liquid biological samples. The most popular technique is Laser Matrix -Assisted Laser Desorption/ionization (MALDI and electrospray ionization. The analytes will be co-crystallized by a matrix and they are transformed into ions by the lasers in MALDI. (Figure 5.7)

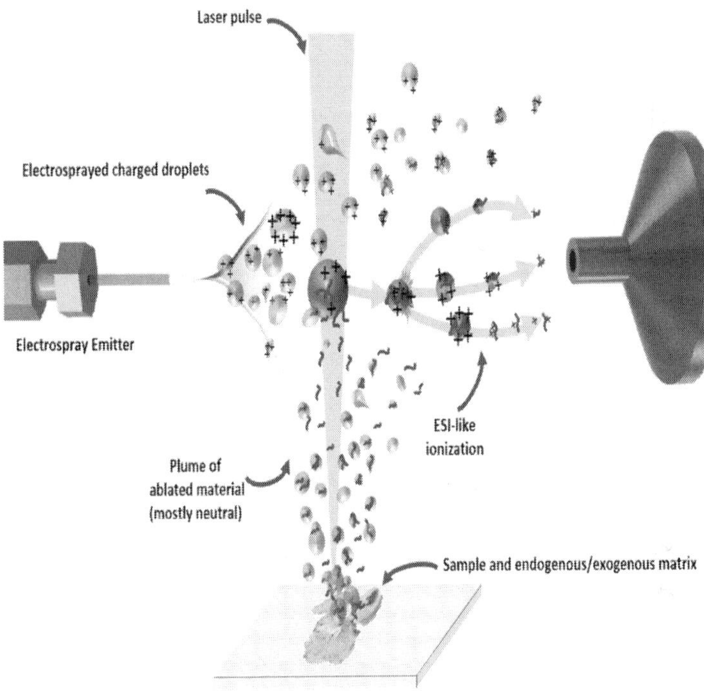

Figure 5.7. MALDI.

Source: https://en.wikipedia.org/wiki/Matrix-assisted_laser_desorption_electrospray_ionization

There are several softwares and databases available and can be used to analyze the spectral data. SEQUEST (Figure 5.8) can help us to identify the mass spectra into protein sequences. This is classified as a first database search program. This program is too slow and it has high sensitivity. It is most beneficial is used to study post-translational modifications (Jimmy et al., 2009).

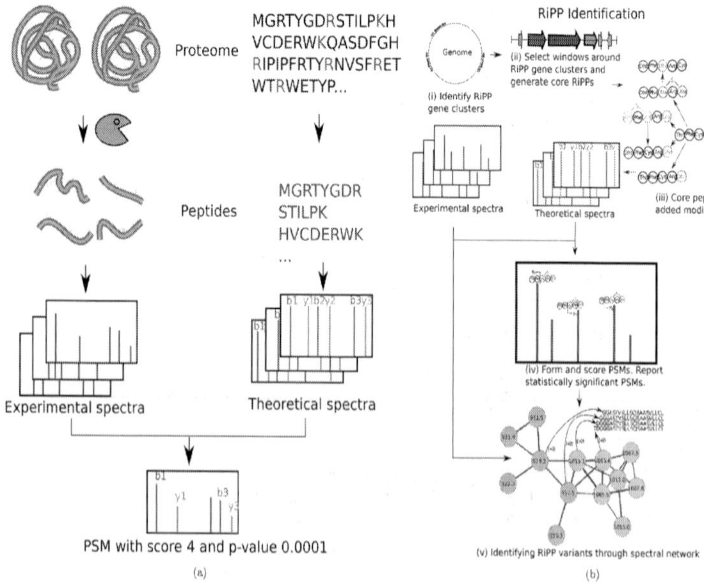

Figure 5.8. SEQUEST.

Source: https://www.researchgate.net/publication/283262317_Dereplication_sequencing_and_identification_of_peptidic_natural_products_From_genome_mining_to_peptidogenomics_to_spectral_networks/figures?lo=1&utm_source=google&utm_medium=organic

Mascot (Figure 5.9) is considered as an efficient search engine that can use data obtained from mass spectrometry. We can identify the proteins by using the primary sequence database (Perkins et al., 1995). Scaffold 3 is another important software that can be used to obtain a high confidence level in case of protein identification. We can use this with MS search engine so that the data can be visualized and validated even with multiple runs.

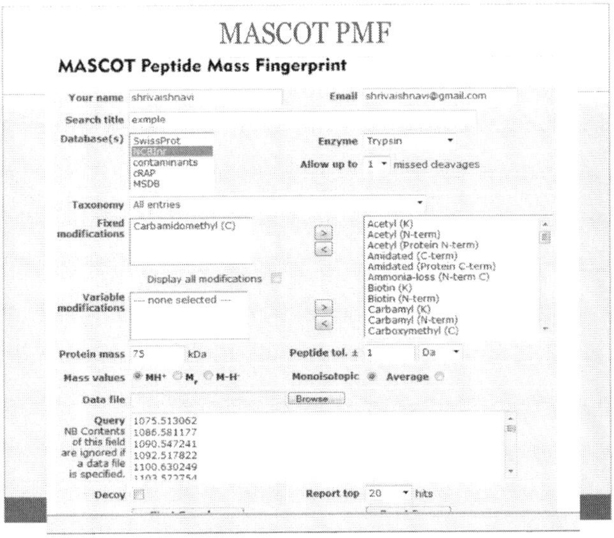

Figure 5.9. Mascot.

Source: https://pt.slideshare.net/vaishnavishankar/mascot/6

5.2 CAPILLARY ELECTROPHORESIS

Jorgenson and Lukacs have described the zone electrophoresis technique and capillary electrophoresis technique (Jorgenson, 1983). Capillary electrophoresis or CE has become popular in peptide analysis. We can use CE as a complementary technique for HPLC. Both CE and HPLC share same physicochemical principles. The CE technique is based on the charge density differences of the compounds. The CE technique can be flexible as we can expect a high rate of resolution. It is very economical as we will use minimal chemicals and sample volumes. We can use limited concentrations of organic solvent or in some cases, we can avoid it. In eukaryotic cells, there are natural peptides that are composed of 21 proteinogenic amino acids. These amino acids are called as L-amino acids. There are several D-amino acids and some amino acids are unusual. Based on the number of amino acids and composition, we can classify peptides. The binding capabilities, hydrophobicity, shape, size and charge can differ in these peptides. The peptide and protein migration in the capillary electrophoresis can be defined by retention and analyte migration in the stationary phase (Bandilla et al., 2006). The analytes can be resolved based on the electric field applied. This can be considered as a function to determine the physicochemical properties. The charge density can be one of the important physicochemical properties. The total charge on the peptides can be due to an addition of charges of the positive and negative groups. The negatively charged groups can be due to carboxyl groups. The side chains of Tyr, Cys,Glu and Asp can also confer the negative groups. The amino groups are the positively charged groups. The side chains of His, Arg and Lys can confer the positive charge on these molecules.

The pH of the background electrolyte can be considered as an important factor in the case of CE technique. This will control the charge present on the peptides. If the peptide mobility is less, then we can expect a high rate of resolution (Castagnola et al., 1996). There are different modes of separation in CE techniques. The capillary zone electrophoresis can be considered as powerful and universal. This technique can be used in the analysis of peptides. Most of the peptides are separated dependent on the size and charge. The mobility of the analytes can be based on the charge densities. The detection of the peptides can be hard in the CE technique. We can use UV radiation in the range of 200 to 220nm. We can detect peptides using UV radiation in the CE technique. The peptide amide bonds can produce the absorbance in the UV region. We can detect the aromatic amino acids like Trp, Tyr and Phe via UV scanning. This can be done by a diode array detector. We can increase the sensitivity by peptide derivatization (Bardelmeijer et al., 1997). Fluorescence can be classified as a detection method with a high rate of sensitivity in identifying the peptides. When UV light is used, we can increase the detection range by at least 100 times. The preparation of samples can be important in the CE process. Most of the samples can be concentrated through electrophoretic stacking techniques, chromatographic techniques, liquid-liquid extraction or solid-liquid extraction.

In the case of CE, we can use the nonpeptide analytes. The sample concentration in the CE technique can be done by electrophoretic preconcentration and chromatographic technique (Sentellas et al., 2002). Any minute change in the size and charge of the peptides can be controlled by CE. Some of the reactions are deamidation, oxidation and hydrolysis. The analysis of peptides using CE can be classified into several categories. The analysis can be done with the peptide maps that are produced from the trypsin digestion. The bioactive peptides that were extracted from the biological samples can be analyzed using CE. The synthetic peptides can be analyzed so that there will be purity control. The peptides can be used as model compounds so that we can study basic aspects of CE. We can determine the pK_a of phosphinic pseudopeptides, enkephalins, di, tri and tetrapeptides using CE. We know the reaction kinetics in angiotensin- converting enzymatic activity, phosphatase activity, kinase and peptide oxidation using CE. This technique can be used in the bioanalysis of physiological peptides. Some of the physiological peptides are neuropeptides, glutathione, amyloid peptthyrotropin releasing hormone, somatostatin, vasopressin, neurotension, angitension II, gonadorelin, cytokines, vasoactive intestinal peptides and enkephalins. The stereoisomer analysis can be done by the CE technique. We can analyze neuropeptide Y, peptide- derived drugs, N-derivatized peptides and di, tri and tetrapeptides using CE. The peptide degradation and stability can be studied using CE technique. Some of the peptides which can be studied are cyclosporin A, LHRH analogs, neuropeptide Y, Asp hexapeptide, Asp tripeptides, goserelin and Insulin. The peptides can be purified with CE. Some of the peptides that can be studies are IgG1 fragment,lecirelin,dalargin, cyclosporine A, tetracosactide, vasopressin, somatostatin, protegrin IB-367, ginseng polypeptide, bradykinin, insulin-like growth factor, hirudin, neuropeptide Y, insulin, cholecystokinin, endorphins and adrenocorticotropic hormone. The peptides can be identified and separated from the peptide complex. Some of the peptides that can be resolved are dynorphin analogs, insulins, enkephalins and other natural and synthetic peptides. We can do peptide mapping using the CE technique. Some of the peptides which can be mapped using CE are β-lactoglobulin, β-casein, interleukins, somatotropin, human tissue plasminogen activator, ovalbumin, granulocyte stimulating factor, erythropoietin and growth hormones. CE with MS can be considered as an effective tool in the discovery of the biomarkers and in the field of proteomics (Herrero et al., 2008).

In peptidomics, both quantitative and qualitative analysis of the peptides in the cells, tissues and organelles can be done (Schulz-Knappe et al., 2005). The polypeptides can be classified as small molecules produced due to protein degradation. The peptides present in the organisms can represent the synthesis; processing followed by degradation of the proteins. Urine and serum can be considered as primary physiological fluids that can be used in peptidomics (Mischak et al., 1997). There are several databases available for studying the peptidome in the urine samples. One of the important applications of the CE technique is the separation of the peptide isomers. Some of the analyses are crucial in monitoring the purity of the synthetic peptides and this testing can be vital in quality control. We can separate the peptide diastereomers by CE techniques. In the CE technique, the efficiency and resolution can be equal to the length of the capillary and there should be a constant electric field. The migration times and efficiency can enhance the resolution and it is based on the square root of the length. Enhancing resolution and efficiency can increase the length of the capillary and there will be increased at the time of analysis. It is important to use long capillaries in analyzing complex mixtures. When the capillary diameter is increased, there will be less resolution and efficiency. The capillaries with small inner diameter permit high voltages and ionic buffer strengths. Due to this, there will be a production of heat. The diameter of the capillary can have minimal effect on the electroosmotic flow. There will be an increase in the sensitivity only if the optical path is enhanced. The capillaries with large-bore can permit the loading of high masses. In the separation of the peptides with CE, there will be fused silica capillaries that are unmodified. In case, the analytes interact with the surface, it is important to rinse with sodium hydroxide solutions. The capillaries that are coated can be employed to lower the interactions between the walls. Hydrophobicity and hydrophilicity of the interior walls can be adjusted. The surface-modified or coated capillaries can modify, stabilize and even eliminate the electroosmotic flow via the generation of a stable but reproducible surface. The coating on the surface can change the selectivity of the separation. The electrolytes with low pH can be combined with positively charged coatings or neutral coatings. In CE techniques, the peak shape, separation efficiency and migration velocity of the solutes can be sensitive to the buffer characteristics. The buffer regulates the migration and ionization of the analytes. It also regulates the magnitude of the EOF or electroosmotic flow. This flow can be directed by the charges of the inner wall of the separation capillary. The conductivity and pH must not be changed even when the migration process is underway. This can be ensured by the buffer systems. The separation selectivity can be optimized by the separation buffer and pH. The basic peptides can be protonated and most of the peptides can migrate in the form of cations. The selectivity is due to the minute differences in the dissociation equilibria. The phosphate buffers can be used at an acidic pH. We can substitute the buffer cation by amines such as triethanolamine and triethylamine so that the peptides can be separated effectively. The amines can have a positive charge at acidic pH. There will be residual charges on the wall of the capillaries.

There will be a suppression of analyte-wall interactions. The buffer capacity can be enough to stabilize pH in the separation process. This buffer capacity can be proportional to the buffer concentration. The buffer capacity can be proportional to the acid and basic buffer species. The analyte stacking effects can be produced using separation buffers of higher concentrations.

The electrolyte concentration can affect the electrical current and so the joule heating. This limits the concentration of buffer. The buffer ions, both the cations and anions can affect the resolution, selectivity, analyte mobility and EOF. The buffer adjustment can increase the separation process. The isoelectric buffers can be classified as special buffers. In these buffers, there will be amphoteric compounds like glutamic acid, aspartic acid, iminodiacetic acid and cysteic acid. In case, the optimization of pH is not enough for obtaining high resolution, we can add buffer so that the differences between the analytes can be achieved. The organic solvents like acetonitrile, 2-propanol, 1-propanol, ethanol and methanol can be used to alter the viscosity of buffer. These solvents can be used to modify the EOF and separation selectivity. If the concentration of the organic solvents is increased, then the electrical current decreases. The separation speed with the organic solvents can be difficult to understand. Acetonitrile can increase EOF and so there will be shortened analysis time. Methanol enhances the migration time of the analytes and trifluoroethanol can be used in the separation of the peptides (Castagnola et al., 1996).

The separation efficiency and resolution can be based on the organic solvents since they affect the dissociation equilibria of the sample volumes. There will be a formation of a pseudo stationary phase due to the addition of detergents. This phase occurs only if the detergents are added beyond the critical micelle concentration or CMC. Sodium dodecyl sulfate or SDS can be used as a surfactant in this process. The capillary surface that is positively charged can be formed due to the addition of cationic surfactants. Some of the cationic surfactants are dodecyltrimethylammonium bromide or DTAB and cetryltrimethylammonium bromide or CTAB. They can reverse the EOF. These detergents can be mixed with cyclodextrins or organic solvents. The development strategy must ensure certain conditions so that the purification process can be a success. Peptide solubility can be important in this process. The analytes must be soluble and they must be stable in different solutions. The capillary dimensions can be critical in this process. It is important to use a fused silica capillary. The inner diameter should be 50 μm and the length can be up to 50cm. These dimensions can help in enhancing the sensitivity of the detection, easy heat dissipation and resolution. To enhance detection sensitivity and mass loading capacity, the capillaries must have larger inner diameters. The diameter can be up to 200μm. The coating of capillaries can be important. The coated capillaries are important in modifying the EOF. The analyte adsorption can be suppressed. We can go for a dynamic and permanent coating. The background electrolytes with low pH can be combined with the neutral permanent coatings or positively charged dynamic coatings. The capillary temperatures should be between 20 to 25°C. The buffer pH can be optimized at the pH range between 2.5 to 4. The buffer concentrations can be optimized with solution concentrations ranging from 50 to 100mM. The selection of buffer additives can be important. The buffer additives must be needed so that there will be maximum selectivity. The organic solvents can enhance the solubility of the peptides via modification of EOF. The soluble polymers and polyamines may lower the hydrophobic interactions between the capillary wall and peptides.

5.3 PROTEIN ELECTROPHORESIS

Proteins can be considered as basic units of cells and they can give us information about the molecular basis of disease, health and others. There are several techniques that can be used to separate and isolate proteins. These techniques can help us understand the functions of the proteins. Electrophoresis can be used as an important tool in this process. It is being used in several proteomic technologies. In protein electrophoresis, there will be the migration of the proteins due to the electric field. The mobility of the proteins can be based on several factors. It is based on the charge on the surface. The pH of the buffer can be important as it affects the net charge present on the surface of the proteins. Any electrophoretic run comprises of sample separations on the gel whilst immersed in the buffer solutions. The gel can be stained and then processed with fluorescent tags or radioactive tags. We can process the proteins using immunoblotting and even peptide characterization can be done. The electrophoretic techniques were described by Tiselius (1938).

The proteins tend to migrate in an electric field and the rate of migration can be based on different factors. The migration and resolution are based on various factors. The pH of the buffer and pI can be considered as important factors. pI can be constant and it is based on the proportion of carbohydrates and aminoacids. The charge carried by the proteins can be defined by the pH. Resolution can be increased by five band method so that the closely linked proteins can be resolved. Other factors that enhance the resolution are the reduced protein interactions with medium, reduction of passive diffusion and rate of migration. The speed at which the protein migrates under an electric field at certain conditions can be considered as an important feature of that protein. The electrophoretic mobility of any protein can be calculated by the following equation

$\mu = d/Et$

Here μ is the electrophoretic mobility, d represents the distance in cm. E represents the electric field in volts/cm. t represents the duration of the electrophoretic run in seconds. With this equation, it is obvious that the electric field strength can be inversely proportional to the length. If we increase the voltage, then the separation will be faster. In some cases, an increase in voltage may generate heat. We can calculate the heat produced whilst the passage of the electrical current. It can be explained by the following equation:

Heat produced = XE^2/A

X represents the specific conductance and E represents the electric field. A is the mechanical heat.

From this equation, it is obvious that the heat can be increased along with the voltage. The heat produced can create a problem with the band formation. Heat can enhance thermal agitation and there will be diffusion of the proteins. Due to diffusion, the bands will be broadened and so the resolution will be reduced. Due to the production of heat, the viscosity will be decreased. The ionic strength can be important in determining the protein bands. The salt concentration in the buffer can reduce the migration of the proteins. The ionic strength may not show any specific effect on the serum proteins. The rate of migration is based on the interactions between the surface charges and buffer ions. The positive charges that are present in the buffer can enclose the albumin molecules that are negatively charged. This can be called a diffuse double layer. It is important to regulate the rate of evaporation during the electrophoretic run (Wieme, 1966). The separation of the bands can be based on the adsorption potential of the molecules on agar. Based on the gel thickness, purity, heat dissipation, voltage and buffer strength, there will be an enhanced number of bands (Laurell, 1973). In the case of the detection of monoclonal gammopathies, it is essential to achieve high resolution (Aguzzi et al.,1992). There are some recommendations made for successful electrophoretic runs (Aguzzi et al., 1986). In the case of serum, it is important to observe that prealbumin band is visible in all serums taken from healthy subjects. It is important to see that α_1-antitrypsin is present and it is in heterozygous condition. We must be able to detect the haptoglobin and α_2-antitrypsin. We need to ensure that the two components of the β-zone called C3 and transferrin are resolved by 2 different bands. It is important to express the results in the form of semiquantitative and qualitative variations for certain proteins. In the case of γ-zone, we must be able to detect the monoclonal components. It is recommended to see that these systems provide enough resolution needed to separate transferrin and C3 proteins so that there will be improved detection of the monoclonal gammopathies (Keren et al., 1994). If the protein concentrations are less than 10mg/dl, then we need to use immunoassay techniques (Wild, 2013). The electrophoretic method is based on the principles proposed by Wieme and it was modified by Johansson (Johansson, 1998). The sample must be added on the agarose surface in a band that is very narrow. In the case of manual systems, we need to eliminate moisture by using filter paper through blotting. This process can be important in the protein diffusion process. The lateral movement of the proteins can be arrested by this process. We need to use a plastic template to ensure that there are uniform slits. These slits are needed in the application of the samples. The slits should be firmly layered onto the gel. This template must be applied on the surface of the gel uniformly so that there should not be any air pockets left. If air pockets are present, the application of the samples will be disturbed.

We need to use at least 3 to 5 μl of the sample. This sample will be placed on each slit. Then, the sample will be permitted to get diffused into the gel. This diffusion time will be about 5 to 7 minutes. It is important to maintain the consistency and attention all over the process. In the case of semi-automated gel systems, we may use devices that have a facility of automatic washing. This is important in the technical process. Whilst dealing with the urine samples, care must be taken so as to ensure that solid matter is restricted. The urine samples can have solid matter comprising of crystals, cells and others. Due to the presence of these solid materials, the wicking process will be disrupted. The wicking of the specimen onto the gel will often be affected due to these solid particles in the samples taken from the subjects. It is important to ensure that the sample does wick accurately. If it is not possible, we should not add the sample onto the gel. There will be no visible bands when the serum is used as a sample in some cases. Post application of the sample, it is important to cool the gel. Most of the gels are cooled by the convection process. We can use the Peltier cooling device to regulate the heat generated in the electrophoretic systems. The buffer concentration in the reservoir can be considered as an important factor in this process. Most of the agarose systems run when the electrical field is set at 20V/cm. It is nothing but adjusting the potential difference so that each 10cm length of the agarose gel receive 200V. The current must be set between 100 to 120 mA. In this setup, we can expect a successful run for 30 minutes or even 50 minutes. Post electrophoretic run, the proteins will be fixed with a fixative. This fixative is called acid fixative. We can use picric acid as a fixative. This fixative can be explosive. Post fixation, the gel is allowed to dry under dry heat for 5 to 10 minutes. In investigating the urine and serum protein samples, we can stain the gels with Amido black. The stains may differ from the systems. Mostly, we can use Amido black and Coomassie Brilliant Blue as stains. Amido Black can offer minimal background between the bands. This is more prominent between the major bands. In this case, the interpretation of both urine and serum samples can be simple. Coomassie Brilliant Blue can be more sensitive than Amido Black. It even stains the minute protein molecules. Whilst using cerebrospinal fluid as a sample, we can use Coomassie Brilliant Blue. In the case of black and white photography, we can use this stain. Since the 1950s, cellulose acetate has been used as a gel media (Kohn, 1957). It is far better than the filter paper. We can achieve uniform porosity with cellulose acetate. We can use various systems that can support cellulosic media and agarose. This can be produced by mixing cellulose with carbonic anhydride in a volatile organic solvent (Destro-Bisol et al., 2000). We can expect 5 band resolution on these membranes. We can use the gelification process to avoid drying. Due to this process, we can get gelled cellulose acetate. This can enhance the resolution and so the number of bands will be increased (Destro-Bisol et al., 2000).

By using the gelled cellulose acetate, the serum proteins can be separated into fractions and we can achieve high resolutions. In the case of agarose gel electrophoresis, the cooling process is not needed. We can perform an immunofixation analysis on these gels. In the case of immunofixation analysis, it is important to use cellogel strips. In order to get a high resolution, we need to equilibrate cellulose acetate with Tris-glycine salicylic acid buffer solution for at least 10 minutes and there should be gelatinized membranes. The cellogel strip must be blotted between 2 filter papers. Then this strip must be transferred into the electrophoretic chamber. We can use a linear sample applicator in this process. It is used to deliver 1μl. We can use this to apply an undiluted serum. The serum can be applied at the cathode end of the gel apparatus. We need to run the samples for at least 25 minutes and the voltage to be maintained must be 300V. These samples should be fixed and then stained. We can use different stains like immunological reagents, gold, Schiff, Nigrosin, Ponceau Red and Coomassie Brilliant Blue. Most of the samples should be destained. The destaining process can be done with methanol-water and acetic acid. In the case of capillary zone electrophoresis or CZE systems, the electropherogram can provide us with fundamental information generated from the gel.

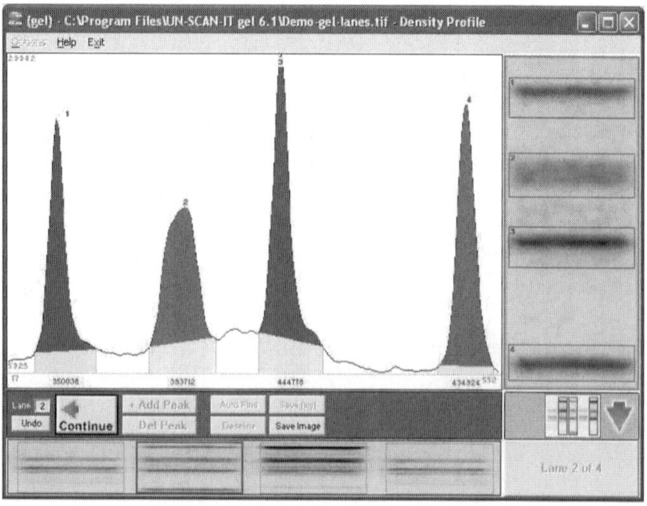

Figure 5.10. Densitometric scanning.

Source: https://www.silkscientific.com/how-it-works-gel-analysis.htm

Densitometric scanning (Figure 5.10) can be classified as an important technique in protein electrophoresis. This technique can give us information that is objective in nature. Care must be ensured that there should be a visual inspection. The densitometric scanning may not detect minute oligoclonal patterns, monoclonal gammopathies and α_1 antitrypsin patterns. It is important to go for visual inspection so that all the fine particulars will be intact. The precision of this technique is poor whilst investigating the α_1 globulin fraction of serum (Kahn et al., 2005). We can classify this technique as a semiquantitative procedure in the case of clinical densitometry. The values obtained can be different and it is based on the electrophoretic system, densitometer, stain and the dilution of the serum. If we use concentrated proteins, most of the binding with the dyes may not be linear. This is true with the monoclonal proteins and albumin (Chang et al., 1997). We can expect linearity if the gels are stained with Amido Black. Amido Black is better than stain Ponceau S in staining protein fractions. We can use silver stains but there might not be linearity as expected. We can use electrophoresis strips that can be stained with Amido Black and they will provide us an estimate of gamma globulin, transferrin and serum albumin. There were reports, stating that the densitometers gave different results. The correlation between the polyclonal IG concentration determined by densitometry and that determined by nephelometry was perfect (Ritchie et al, 1983).

5.4 IMMUNOSELECTION TECHNIQUES

The serum protein electrophoresis may not identify the M-protein and it is important to go for immunochemical methods that are linked with electrophoresis. M-protein can be identified by immunofixation electrophoresis or IFE. Immunosubtraction (Figure 5.11) or ISUB is possible with CZE. Immunoelectrophoresis or IEP is being used by many labs. Immunoselection or ISEL can give us more information than conventional techniques. The fundamental precipitin reaction is the same for almost all these techniques. In the case of the immunoprecipitation technique, there will be an interaction between the antigens and antibodies. This interaction can occur in the liquid matrix or gel matrix. In this matrix, the molecules can be free so that there will be diffusion. The main cause of precipitation can be the multivalent nature of the antigens and antibodies (Davis, 2001). In chemical reactions, if a substance P is mixed with Q, then we can expect R. R is considered as a product.

$$P + Q \rightarrow R$$

If we increase P substance and keep Q as constant, then the precipitate R may get accumulated to a certain point. After that point, the precipitation process will stop even if we add more P. This shows that all the P molecules were combined with Q. The antigen and antibody reactions can differ with the normal chemical reactions.

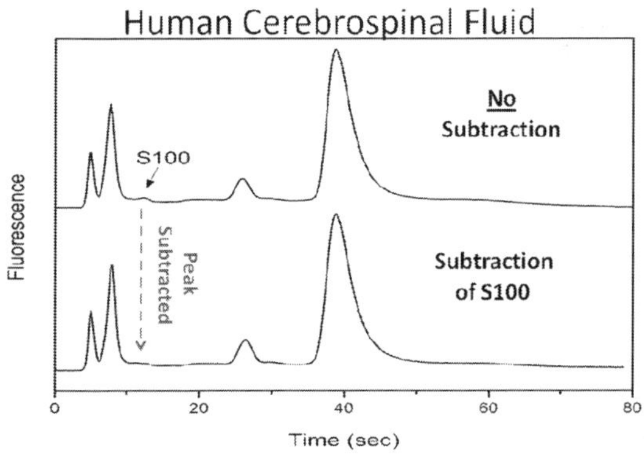

Figure 5.11. Immunosubtraction.

Source: https://www.researchgate.net/publication/267961277_Assessing_the_traumatic_brain_injury_markers_S100_and_C-reactive_protein_in_human_cerebrospinal_fluid_via_microfluidic_immunosubtraction/figures?lo=1&utm_source=google&utm_medium=organic

These interactions can be complex and we need to consider more variables. Most of the antigen molecules can have epitopes or surface determinants. The antibodies interact with these epitopes that lead to their binding. Each epitope can produce B lymphocytes of variant clones. There will be antibody-secreting plasma cell differentiation. Antibodies that are produced by a clone can be different and this variability is specific to the epitope. The tendency of an antibody to bind an epitope can be called its affinity. This binding can be classified into 4 types. These interactions in the binding process can be classified as non-covalent interactions. They are called coulombic forces, H-bonding, hydrophobic bonding and van der Waals forces. The coulombic forces can be due to attraction between the groups that are oppositely charged. The opposite charges can be due to COO^- and NH_3^+ groups. The H-bonding is due to the binding of a H atom with an electronegative oxygen atom. The hydrophobic bonding can be due to the self-association of apolar groups. The van der Waals forces can be between the outer electrons. These forces are classified as weak forces. If the distance between the antigen and antibody decreases, these forces will get enhanced. All these interactions are weaker than covalent bonding interactions. This shows us that the antigen and antibody interactions are reversible. The interaction strength of an antibody with an epitope can be based on these four forces of attraction. It is important to know the concentrations of both antibody and antigen in optimizing ISEL, IEP, ISUB and IFE techniques so that the anticipated errors can be addressed quickly.

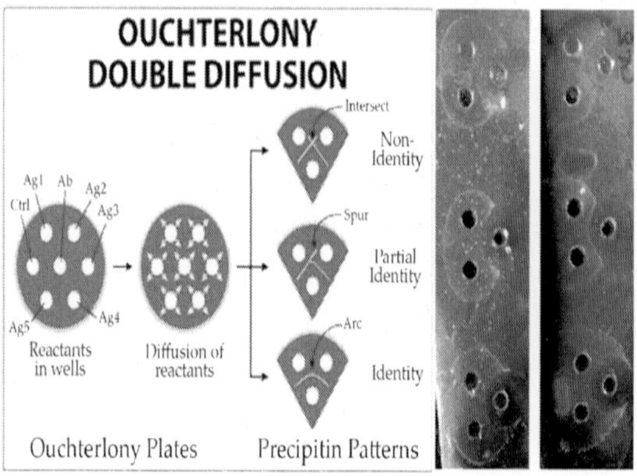

Figure 5.12. Ouchterlony technique.

Source: https://microbenotes.com/ouchterlony-double-immunodiffusion-technique/

The double diffusion occurring in two directions can be called as Ouchterlony technique (figure 5.12). Ouchterlony prepared a simple method to apply the immunoprecipitation reaction to estimate the antibody reactivity against an unknown antigen (Ouchterlony, 2015). We need to cut the wells in an agarose gel accurately. We need to inject an antigen in one well and the antibody in another well. Most of the molecules follow radial diffusion. This process will happen out of the wells. These molecules diffuse way form the mid-point of the well. The concentration of the molecules will get decreased. This decrease can follow a logarithmic pattern. We can identify the precipitin band at a point where there is an equivalent concentration. In case, a precipitin forms close to the antigen well, then we can infer that the antibodies are more concentrated than the antigens. This is because the high number of antibodies can travel further than antigens. Due to this migration, the antibodies can get diluted even before the precipitation. There are some other factors such as size and interactions that can affect the gels. Ouchterlony used the known antigen and antibody that can estimate the unknown Ag.The development of the immune precipitin reaction led to the development of immunoelectrophoresis.

This technique combines immunological techniques and electrophoretic techniques. Immunoelectrophoresis was considered as an essential technique in most of the labs for both clinical and research purposes. In this technique, the serum is injected into the wells of the gel. The gel is made up of agarose. Due to electrophoresis, there will be a separation of the serum proteins. Once the electrophoresis is completed, the gel will be removed from the apparatus. The troughs formed will be filled with the antisera. We can use antisera to different compounds.

These compounds may include the antipentavalent human IgG. This can react with 3 light and heavy chain classes. The other troughs can be filled with antisera that are monospecific in nature. This antiserum can be used against λ and κ chains, IgM, IgA and IgG. In order to compare the migration patterns, it is important to ensure that the antisera are placed in the apparatus are correct. In order to ensure that, we need to use the control serum and sample on the strip. Most of the antisera can diffuse into the gel. The diffusion is quite slow. We can observe a radial fashion of diffusion. Mostly, the precipitin lines get formed only if the antigens and antibodies are at equivalence.

The precipitin lines can be arc -shaped. We can identify monoclonal proteins easily with IEP. This can be done by comparing the serum sample with controls. The immunoelectrophoresis can be used in the detection of the monoclonal proteins that are linked with Waldenstrom macroglobulinemia and multiple myeloma. If monoclonal proteins are in large quantities, we can use IEP to detect the diffusion effect.

In the clinical laboratory, immunofixation electrophoresis can be considered as an important development. This technique can be run for about 4 hours. In the case of immunofixation, the sample serum is diluted and it is applied onto the gel. The lane that is used for comparison can be fixed with the aid of acid. In the case of pentavalent immunoglobulin lanes, the sample can be overlaid with antiserum. Once the antisera are removed, we need to wash gels. The gels can be stained with Amido black and Coomassie Brilliant Blue. In some cases, we can use protein dyes. Silver staining can be done to improve sensitivity. The immunosubtraction technique was used in gel-based electrophoresis. This technique can be used to subtract out many proteins. They are fibrinogen, C3, β-lipoprotein, transferrin, fibronectin, haptoglobin etc. The immunoselection techniques are based on the design of the immunosubtraction. We can use this technique to investigate the heavy chain disease. To investigate the heavy chains, a modified ISEL procedure has been used (Sun et al., 1994).

6

Nanotechnology in Chromatography and Electrophoresis

CONTENTS

6.1 INTRODUCTION

The separation techniques can be considered as a key in agricultural, geochemical, environmental, pharmaceutical, chemical, biotechnological and biological disciplines (Cheng, 2004). Most of the drugs are present in the biological systems in low concentrations. The pharmacodynamic and pharmacokinetics studies can be based on analytical techniques.

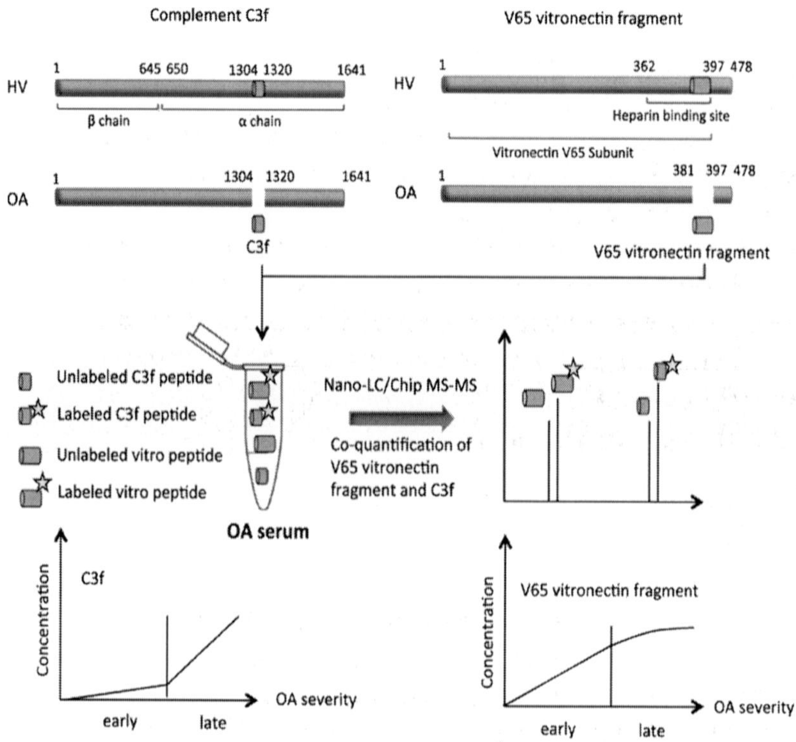

Figure 6.1. Nanoliquid chromatography.

Source: https://ars.els-cdn.com/content/image/1-s2.0-S0039914017303740-fx1_lrg.jpg

These techniques can be used in detecting the pharmaceuticals and drugs even at nano levels. Some of the drugs can have side effects at even femtogram and nanogram levels (Tse, 2006). The concentration of proteins, antibodies, DNA and RNA can be low in the biological systems. There are some xenobiotics like phenols, plasticizers, polybrominated diphenyl ethers or PBDEs, polychlorinated biphenyls or PCBs, polynuclear aromatic hydrocarbons or PAHs and pesticides that can be toxic to the ecosystem (Gaston, 1979). These pollutants can cause diseases and many side effects in humans. At this juncture, it is important to use advanced techniques to detect xenobiotics, pharmaceuticals and drug detection. The detection techniques must be able to resolve the analytes at nanolevels. Nanoanalysis can be important to detect the chemical species at nanogram levels. Nanolevel detection is continuously increasing these days. The nanoanalysis has become a common technique among several scientists.

Nanoliquid chromatography (Figure 6.1) was introduced by Karlsson and Novotny (Karlsson et al., 2010). Nanolevel investigation is needed when a minute concentration of samples is present. Detection can be considered as an important aspect in Nanochromatography. We can define Nanochromatography as a chromatographic technique that involves nanoliter sample volumes and the mobile phase flow will be nanoml/min. The Nanochromatography technique can be run on a chip. This can be called as lab-on-chip chromatography. This can be called Nanochromatography or NC. If we use nanotechnology in liquid chromatography, we call it nanoliquid chromatography or NLC. In the same way, the application of nanotechnology in capillary electrophoresis can be called as nanocapillary electrophoresis or NCE.

6.2 NC AND NCE

Both NCE and NLC can be considered important techniques these days. These tools can be effective if the analytes are in minimal concentrations. The internal diameter of the columns can be used in NCE and NLC. The detectors used in these techniques may need minimal flow rates. The detectors that may be used are mass spectroscopy and electrospray liquid chromatography. We can expect high sensitivity with these columns. The sensitivity is due to dispersion features. Microfluidic systems are gaining importance in the life sciences field. The principles guiding both NCE and NLC are almost the same as CE and LC. The main difference is that there will be miniaturization. The miniaturization can be advantageous on several fronts. There will be appropriate hyphenation of the detectors. This may need a mobile phase flow. We can expect the mass separation of chips simultaneously. There will be a low pressure on the mobile phase. There will be a narrow peak width visible on the electropherogram and chromatogram. Due to miniaturization, there will be high sensitivity and speed. We can get reproducibility easily using the nanolevel applications. We can expect a high range of portability and this can be achieved by reducing the system size. This can be classified as cost effective. In this process, there will be less consumption of electricity and solvent. The operational time will be much reduced. There will be reduced production of waste. The solvent consumption will be reduced.

NCE can be considered as the latest development in the separation technology. It can be used in genomics and proteomics (Khandurina et al., 2002). This technique can be used for sample volumes that can otherwise be hard to separate by NLC. The separation logic is different. NCE has low detecting limits and this will help us in characterizing minute quantities of the substances present in the samples. Most of the enzymatic reactions can be allowed inside the capillary so that the analytical purposes will be fulfilled. The microdevices for NCE and NLC can be fabricated and regulated by integrated circuits. There is a need for special materials and processes to be used in the micromachining process (He, 2005). (Figure 6.2)

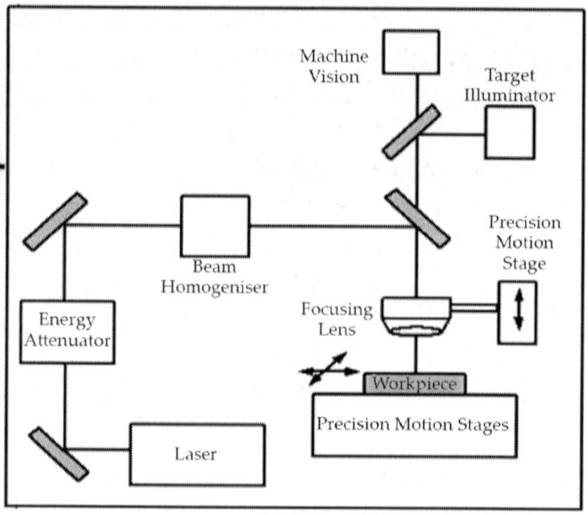

Figure 6.2. Micromachining.

Source: https://slideplayer.com/slide/10887704/

μ-TAS or Microtonal Analysis Systems can be classified as a fastest developing technique in the fields of biotechnology, analytical chemistry, material science and microfluidics.

We can study picoliters, nano and microliters of fluids under the branch called microfluidics. There will be both integration and miniaturization of the functionality of both biology and chemistry lab on 'Lab-on-a-Chip' technology. We can do detection, separation, reactions and preparations on this chip. µ-TAS involves reduced sample reagents, low power and size. We can reduce operational costs and manufacture needs. With this technology, there will be automation, mass sensitivity, throughput and speed. Microfluidics can be considered as an important aspect in both NCE and NLC. The microfluidic system that is miniaturized can be used to automate the functions so as to transform the chemical information onto the electronic data. Terry et al developed µ-TAS (Terry et al., 2015). This device was developed for gas chromatography. This technology was used in several disciplines (Reyes et al., 2011). In the 1970s, these microdevices were part of the silicon microprocessors industry.

The nanoliquid chromatography or nano-LC concept was proposed by Karlsson and Novotny (Karlsson, 2011). This technique was used with the traditional liquid chromatography. The nano detection was done by modifying the LC systems. If the diameter of the column is less than 100µm, then we can call such a chromatographic system as nanochromatographic systems. The integration and miniaturization of electronic devices can result in the production of MEMS or microelectromechanical systems. We can do a complete analysis in just seconds. These machines demand a high range of compactness. This property can allow them to test the environmental samples. The chip technology holds good with NCE than conventional methods. This is because of the difficulties in integrating the mechanical valves, injectors and chip pumps and problems with the flow control (De Mello, 2013). It is hard to fit a stationary phase in the NLC. The sealing of the microchannels can be a problem. Manz et al demonstrated NCE with silicon + glass chips (Manz et al, 1985).

The separation can be characterized by calculating the retention times, resolution factors, separation and capacity. It is important to know the symbols and terms that are used in the chromatography process. The separation mechanism in NCE can be dependent on the electrophoretic mobility of the species. The migration of the chemical species can be regulated by the intrinsic electrophoretic mobility and electroosmotic mobility. The capillary electrophoretic methods and chromatographic methods are important in the analyses of the pollutants, pharmaceuticals and drugs in the environmental matrices and biological specimens. First, the samples are collected by appropriate techniques. The samples can be classified as solid samples and liquid samples. The solid samples will be ground and homogenized. The liquid samples are purified and concentrated. The solid samples or liquid samples are subjected to microextraction. The microextraction can be classified into μ-liquid extraction, μ-solid phase extraction and μ-chromatographic extraction. After microextraction, there will be a purification test. Post purification, we need to analyse the micro extracts. Finally, we need to go to the detection process. These techniques involve nano and micro amounts of the samples and so they are difficult. It is important to ensure that an expert is operating the NCE and NLC. There is a need for a fail- proof protocol so that there will be successful electrophoretic and chromatographic run.

6.3 FABRICATION

The integrated circuits or IC have helped the development of microfabrication technologies. These integrated circuits can be used in the development of biomedical and microchemical systems or Bio-MEMS. (Figure 6.3)

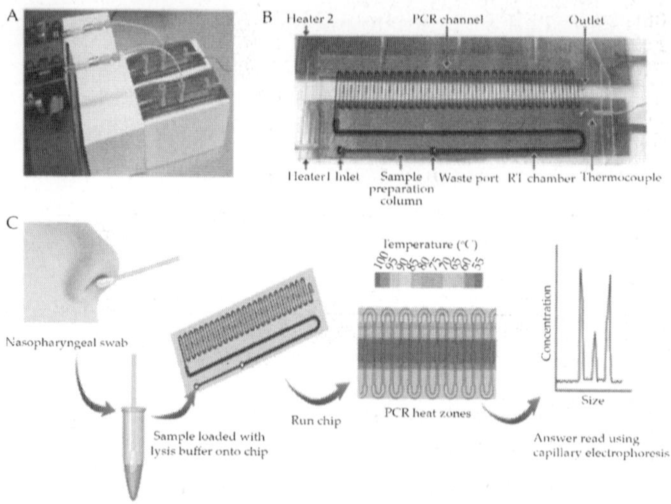

Figure 6.3. Bio-MEMS.

Source: https://en.wikipedia.org/wiki/BioEMS#/media/File:Continuous_flow_PCR_schematic.png

They can be useful in microfluidic devices, genomics, biotelemetry, diagnostics, DNA testing, biochips etc. The flow rate and the fluid movement can be regulated by electroosmosis, microvalves and micropumps. It is important to install electrodes and microcomponents in these systems.

There are many MEMS devices like sensors, valves, channels, rotors, etc that are useful in these systems. The fabrication of the microfluidic chip was described by Chun et al (Chun, 2010). The materials that are employed in the microchips can define the selectivity and efficiency of the nanoliquid chromatography or NC and NCE.

The chips can be made from plastic as it is cost-effective. The plastic chips can remove contamination and they are disposable. Parylene can be used as an appropriate substance in the case of nano applications. It can be used in microfabrication techniques.

This can be used in capillary electrophoretic channels, electrospray channels, microvalves and microchannels etc (Licklider et al., 2002). The fabrication of microchips can be done through several steps.

We can use laser technology and lithography in the fabrication process. We can use machining and finishing in the fabrication process. Most of these processes can be executed by ion implantation, patterning, etching, cleaning, wafer fabrication, packaging and material selection.

The microfabricated devices must be formed on the support substrate. Microchips can be perforated by a process called photolithography.

These openings will be at a range from micometer scale to nanometer scale. We can perform the etching process by exposing the microchip to an etching agent.

The etching agent can be plasma or acids. The surface preparation or wafer cleaning can be done by mechanical planarization, ion implantation and thermal diffusion.

Two of the microchips can be attached to produce microchannels. The fabrication process can be based on applications and substrates. The fabrication process of microfluidic devices can be the same as the micromachining process in the case of microelectronics.

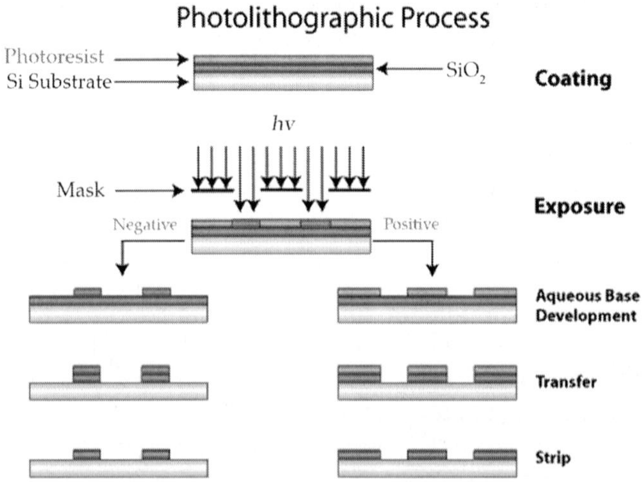

Figure 6.4. Photolithography.

Source: https://www.researchgate.net/publication/230583356_Top-Down_approach_to_nanotechnology_for_cell-on-chip_applications/figures?lo=1

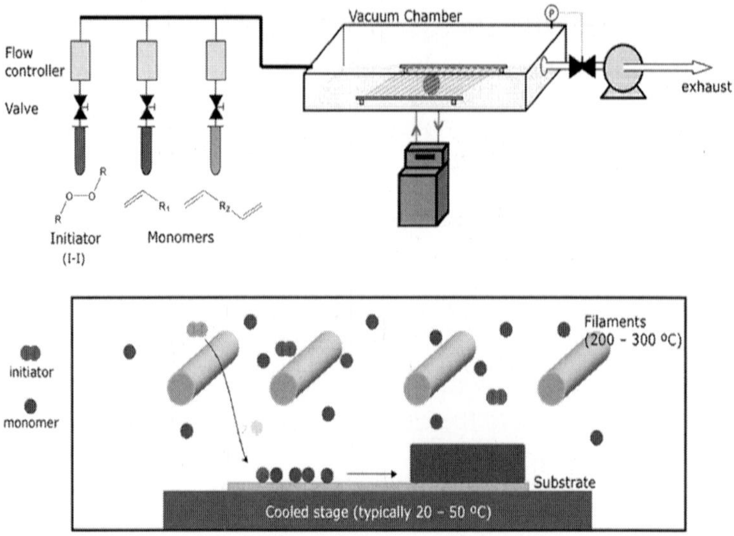

Figure 6.5. Chemical vapor deposition.

Source:http://news.mit.edu/2015/explained-chemical-vapor-deposition-0619

Photolithography (Figure 6.4) can be considered as an important technique in the fabrication of the microfluidic chips (Yang et al., 2012). Chemical vapor deposition (Figure 6.5) and reactive ion etching can be also considered important in this process (Jacobson et al., 1982). The multi-photon and laser ablation can be classified as novel techniques used in the integration process. We can use dry plasma etching, wet chemical etching, sputtering, evaporation and chemical vapor deposition in the fabrication process (Wolf et al., 2006). In the fabrication of MEMS, we can use 3-D stereolithography, (Figure 6.6) electroplating (Figure 6.7), wafer bonding, electroforming and x-ray lithography (Figure 6.8) (Madou, 2004). Polymer MEMS can be classified as an important technique in soft lithography (Xia et al., 2010). One or more techniques are united to build a device through a process called surface micromachining and micromachining (Kovacs et al., 1998). In order to integrate the transducers and optical components to the microfluidic devices, the injection molding and embossing methods are being used (Qi et al., 2007). These techniques can be used to mold the thermoplastics. CMOS or Complementary metal oxide semiconductor processing was used to produce microfluidic chips.

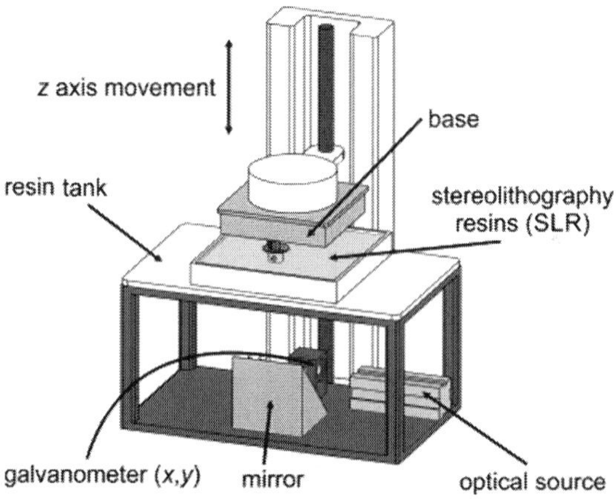

Figure 6.6. 3-D stereolithography.

*Source:*https://www.sciencedirect.com/science/article/pii/S1359835X16301725

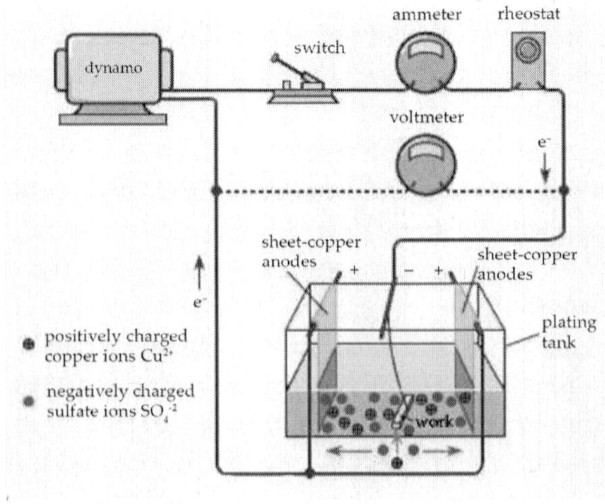

Figure 6.7. Electroplating.

Source: https://www.britannica.com/technology/electroplating

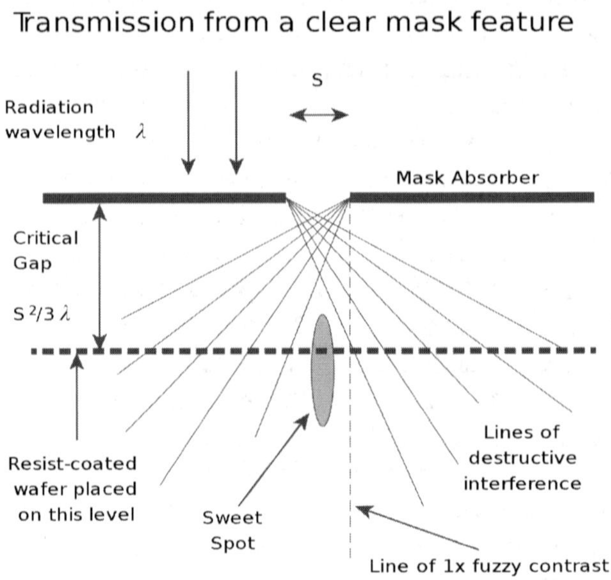

Figure 6.8. X-ray lithography.

Source: https://en.wikipedia.org/wiki/X-ray_lithography

These chips were integrated with the transducers in silicon and glass materials. Glass is considered effective in the fabrication of the microchips. The fabrication of microchips made up of glass was done using photolithographic techniques and etching procedures (Rodriguez et al., 2000). Microscope glass was used in the production of microchips (Fang et al., 2007). Tsai et al (2018) demonstrated the plasma polymerization technique. In this technique, there will be a modification of glass chip in isoelectric focusing. Microfabrication was done using microscopic glass slides (Lee et al., 2004). The fabrication process includes BOE etching followed by electron beam evaporation. The electron beam evaporation is with gold or chromium. Then, there will be gold or chromium etching. This step is followed by cover drilling. Finally, there will be alignment and bonding (Lee et al., 2004).

Quartz can be considered for the microchip fabrication process. Ujiie et al have used quartz chips for NCE (Ujiie et al., 2000). The fabrication of the quartz chips can follow a sequence of steps. First, photolithography is used. Then there will be etching followed by stripping. We need to use sputter deposition and photolithography. Etching and sawing must be done. Finally, there will be mounting and wire bonding (Buttgenbach et al., 2008). Silica chips can be considered as an important material for microchip fabrication. Licklider et al demonstrated the electrospray chips (Licklider et al., 2000). Polymer chips were used in the production of the microchips. The plastic materials are being used in the microfabrication process. The plastic materials offer advantages such as disposability, inexpensiveness and molding. Roberts et al demonstrated a laser ablation technique to produce microchannels in the plastic chips. In this technique, there is no need for fabrication (Pethig et al., 2012). The channels produced will be sealed via a lamination technique. We can use a UV excimer laser to ignite the microchannels onto the substrate. This process is controlled by computer programming. We can use substrates like polyethylene terephthalate or PET and polystyrene or PS. This method is very useful as it has limited effects on the surface of the substrates (Xu et al., 2002). The PCR microchip or PCR microdevices can be studied to know the processes like fabrication, bonding and sealing. This PCR microfluidics can be considered as an important biological tool. These tools can be used to investigate replication and cyclic patterns in PCR. In general, PCR takes about 1 to 2 hours. The chip-based PCR can be used to replicate the DNA strands in seconds (Northrup et al., 1999). This can be considered as a promising technique in proteomic and genomic research. We can produce PCR chip devices using the silicon micromachining technique. We can use chemical vapor deposition, ion etching, electrochemical etching and photolithography in this technique. In conventional chromatography, there will be two phases called the stationary phase and the mobile phase. The polymeric substances can be used to modify the surface. There will be microchannel packing in NLC and NCE. Belder et al demonstrated the vinyl alcohol coated microfluidic devices in NCE (Belder et al., 2002). In this technique, there will be reduced electroosmotic flow. Due to this, the separation will be good if amines are labeled. Pumera et al (2011) demonstrated that the gold particles can enhance the efficiency and selectivity in the NCE process. The walls of the microchannels can be coated with citrate stabilized gold nanoparticles.

Silica gel is considered as a stationary phase in LC and CE. This gel can be used to pack and coat the microchannels. The columns can be made up of monolithic silica. It is used in place of particle-based columns. Most of the columns are biporous in nature. There will be large pores which can be about 2μm. Due to this, there will be a high flow rate. We can yield reduced back pressure. We can analyse even with a high flow rate without comprising the separation rate. The monolithic stationary phases can have several advantages. There will be an enhanced surface area. The surface chemistry can be easily regulated. We can prepare these columns easily. This is due to free radical polymerization within the microdevice. We can control the pore size, surface area and porosity by regulating the concentration of the monomer solution. Adjusting the polymerization conditions can also alter these factors.

The length, size and shape of the microchannels can be important to yield separation in NCE and NLC. Different types of microchannels are designed and fabricated. In NLP microchip, there will be an access hole on the wafer. The column size can be based on the requirement. The columns and capillaries are needed in NCE and NLC. It is important to know that there is a binding of 2 chips so that the microchannels are formed. It is important to ensure that the chips are sealed and there is no leakage. We can use mechanical holding, gluing and pressurizing process to ensure no leakage.

6.4 INSTRUMENTATION

The reproducibility, detection, speed, robustness, selectivity and efficiency of all the techniques can be important. Almost all parts of the instrument are important in the nanolevel analyses. The nanoanalysis can be performed by capillary electrophoresis and chromatography. It is classified as a sophisticated separation technology. In the case of Nanochromatography, we need to consider columns, recorders, detectors, injectors, pumps, tubings and mobile phase reservoirs. The nano detection in samples that have trace amounts of ingredients can be difficult. It is important to know the details about the sample preparation devices so that the desired task can be achieved. Nanoanalyses can be done by capillary electrophoresis, gas chromatography and liquid chromatography. Several studies explain the nanoliquid chromatographic instruments. These studies are mainly methodological, technical and theoretical in nature. Most of the nanolevel separations are due to nano-HPLC. Micellar electrokinetic chromatography and capillary electrochromatography can be considered as important. Szekely and Freitag have described the chip-based system that has a monolithic stationary phase (Szekely et al., 2001). There are no advanced devices that can deal with the mobile phases in case of nano-HPLC. The solvent containers must be free from contamination and it should be airtight. We can use helium gas as a sparging device. It can be used for degasification of the solvents. It can reduce the background noise in the case of UV detection. Each reservoir should be installed with a valve for shutoff so that there will be an effective consumption of helium. The mobile phase can be considered as an important task in the liquid chromatography. Most of the solvents that are to be used should be of HPLC grade. All these solvents must be degassed. Most of the nano-HPLC systems are equipped with degassers in the pumping units. Propanol and ethanol can be used in nano-HPLC. The accuracy of flow and reproducibility of the mobile phases can be regulated by microchip flow sensor technology. These sensors can be important to get accurate and precise flow rates. In the case of nano-HPLC, we need high grade tubings and they are made up of steel. We can use other materials that are made up of polyether ether ketone (PEEK). The interior surface must be smooth so that there will be laminar flow at constant rates in the mobile phase. The alignment of bores and tubing concentricity must be more effective so that we can get nanoscale separation. In order to get laminar flow at constant rates, the connection between the tubings should have less connection gaps.

The nano-HPLC require reliable pumps. They must provide flow rate with negligible variations. The pumps should deliver flow rates up to 4000nL/ min. These pumps exhibit a flow gradient. We can expect high rates of reproducibility even at low flow rates. The flow rate can be obtained due to nano splitting technique. We can use a non-invasive microchip nanoflow sensor for this purpose. The flow splitting pumps can produce enhanced flow rates and so there will be faster gradient equilibration times. Most of the electroosmotic pumps do not have mechanical parts and so there will be a continuous electroosmotic flow. Most of the electroosmotic pumps can be feasible and attractive (Manz et al., 1990). Sample injection can be considered as an important technique in nano-HPLC. It needs flexibility, reproducibility and quality. We need to use low dispersion method in nano-LC. We can get sharp peaks and so there will be a high resolution. The packaging and dimensions of the microchips can determine the efficiency, ruggedness, reproducibility of separation. The chip is considered as an important part in the nano-HPLC systems. The dimension of the chip and the package materials can be based on the experimental conditions and applications thereof. It is important to ensure that an expert selects the dimensions of the chip and the packing materials that are to be used.

If the chip selection is done with the utmost care, we can reduce the costs and time so that there will be increased productivity. The analytical results will be fast and reliable. Some of the other factors that can lower the performance of the columns are laminar flow disruption, non-specific interactions and void volumes. It is important to use PEEksil, integrated connection capillaries and packing materials of high quality so that the nanoscale separation can be achieved. Most of the surface of the materials can be deactivated. The silica gels can have silanol groups and they must be deactivated. In the case of nano-HPLC, the chips can be optimized by reducing the dead volumes. Most of the steel fibres can be gold plated. We can covalently deactivate them. The inner diameter and length of the column can be classified as important features needed in the production of the columns on the microchips. We can measure the separation capacity of the columns in the form of plate numbers. Mostly, the length of the column can be proportional to the number of plates. The average length of these columns can be up to 5cm. The narrow bore columns can be used in trace analysis. In this analysis, we may need less mobile phase and sample. On a chip, most of the columns comprise of rectangular cross-sections. Nano-HPLC can be called as nano-scale HPLC and Nano bore HPLC (He, 2016). The internal diameter of these chips can be up to 100μm.

Most of the nanoanalyses can be sensitive, reproducible and efficient. The recorder has a major role in the analysis part. It can be used for printing purposes especially chromatograms. Most of the nano-HPLC instruments are automated by computers. The trace analyses in the matrices can be important as there will be unknown impurities and so much interference will be expected in detection and separation. Most of the impurities present in the environmental matrix and biological matrix can render the nanoanalysis process complicated. The sample preparation process need to be optimized. The solid phase extraction was done online by Veuthey et al (Veuthey et al., 2018). This extraction was useful in the nanoanalysis of drugs. The sample is injected onto the extraction support. The capillary electrophoresis or CE can be classified as a separation technique that comprises of nanoliter samples. It is fused with silica. The length of the capillary will range from 30 to 100 cm and the diameter will be up to 100μm. The detection limit of this configuration can be stated as microlevel. It can be commonly used in the micrometer levels. The chip based capillary electrophoresis has many advantages. Some of the advantages are high separation efficiency, reduced sample volumes and high speed (Manz et al., 1992). The chip-based CE can be used in the analytical sciences, biological, clinical and chemical sciences. The CE chips can be employed in detection, separation and preparation of small samples (Nakanishi et al., 2000). The microchannels can be produced from quartz glass. This can be prepared by photolithography and chemical etching. The background electrolytes can be considered as an important factor. As of now, there are no special devices available to be used as background electrolytes or BGE in the NCE process. In this technique, there will be a small vessel that can be etched on the chip. Then it is filled with BGE. The sample injection in NCE process can be important to reproduce results. The sample injection process in NCE process is important (Futterer et al., 2005). Several electrokinetic techniques are based on the surface properties of the walls. This can result in producing a bias in different species (Alarie et al, 2001). This can happen only if the surface properties were altered by the samples injected. The detectors present in the capillary electrophoresis can be considered as an important component in the nanoanalyses.

The target for most of the clinicians, scientists and academicians is to get nano detection limits in both environmental and biological matrices. The detector plays an important role in capillary and chromatographic electrophoresis. The mass spectroscopy can be important in both NCE and NC. These techniques can be used to detect nano levels of detection. The MS detectors can be used to enhance sensitivity and gain structural information. With MS we can investigate compounds. NCE-MS and MLC-MS can be compatible as there will be almost the same flow rates. The flow rates can be up to 600nL/ min (Edwards, 2006). The microchips can be used to improve mass detection and sample preparation. NCE-ESI-MS constitute a key innovation in the genomics field (Willems et al., 2005). It can be used in identifying biomarker and glycomics (Amon, 2000). Fluorescent detectors can be used as important tools in the detection of ingredients at nanoscales. These detectors can be useful in both CE and nano-LC (Toyohide et al., 2010). These detectors can offer high sensitivity and selectivity. The electrochemical detectors can be useful in NCE and NLC. There will be an increase in sensitivity. Electrode alignment can be important so as to achieve reproducible measurements. The alignment of the electrodes can be considered as a hard task (Mellors et al., 2016). The separation channels can be shorter than the normal CE and so the analysis time will be reduced to seconds.

High Performance Liquid Chromatography can be a modality of LC and it can separate the samples with high resolution. The pre-treatment of sample and separation of analytes can be done in a short time in case of nano-HPLC. The nano-HPLC can be used both identifications of separation of molecules in environmental sciences, pharmaceuticals and proteomics. Sample preparation can be considered as an important task in the HPLC process. Nano-HPLC can be used in the analysis of compounds in biological samples. The selection of mobile phase and chip can be based on the molecules that are to be tested. This technique has many applications in proteomics. The proteomics field involves protein sequencing, peptide mapping and post-translational protein modification. It is considered as an important technique when traces of samples are available. Song et al determined histamine in brain tissues of a rat with Nano-HPLC-MS (Song et al., 2012). We can use water as a mobile phase. We can use heptanesulfonic acid, formic acid and acetonitrile as ion pairing reagents. 2D nano HPLC can be done with ion exchange chromatography to produce peptide fractions. Huang et al described the nano-HPLC-MS to know the proteomic profile but it is label free (Huang et al., 2011). Srbek et al (2007) used nano-HPLC-ESI-MS for identification and separation of protein mixtures extracted from the rice seeds. Gaspari et al explained the identification of the proteins even at the nanogram levels (Gaspari et al., 2017). We can use Nano-HPLC in drug discovery and quality control too (Lee et al., 2000). Due to the minute amounts of the pollutant samples present in the environment, we can use nano-HPLC techniques. The Nano-HPLC technique can be considered as an important tool in monitoring the environmental pollutants, especially in the industrial areas. Several assays can determine public safety and health. Nano HPLC can be classified as a sensitive instrument and expertise is needed in both mobile phase preparation and sample preparation.

7

Advances in Protein purification Technology

CONTENTS

7.1 INTRODUCTION

Most of the protein purification techniques utilize agarose beads. These beads have all the affinity functionalities like antibodies, glutathione and IMAC. The selection of the functional groups is based on the desired protein that is to be purified. Agarose can be used in the protein purification matrix for several reasons. This chemical can exhibit a low rate of binding to the proteins. This will enhance the purity of the protein fraction that is eluted. The agarose molecules can be hydrophilic in nature. This chemical can offer a three-dimensional mesh that can permit the proteins to diffuse into. Most of the biomolecules can interact with the outer shell of the agarose molecules. This is more prominent at the first micrometers of the shell. There will be a large surface for interaction between the desired proteins and functional groups. These features can enhance the level of protein purification. The agarose matrices can be non-magnetic in nature and some matrices can be linked with magnetite and so they will be magnetic in nature. Most of the non-magnetic agarose beads can be large. The diameter can be up to 150μm. The back-pressure will be reduced in the case of chromatography. The beads that are magnetised can be up to 40 μm in diameter. The minute beads can offer us maximum surface area. We can expect high rates of interaction. It is better to use 10 μm if ferrimagnetic beads are to be used. The primary difference between the magnetic and non-magnetic agarose beads can be determined by the surrounding medium. In the primary purification steps, the medium to be used can be cell lysate or bacterial and in some cases, we can use physiological buffer and cell culture medium. In the elution and washing steps, the medium must contain buffers that have different properties such as pH and salt concentrations. These buffers induce binding and there will be the elimination of the contaminants. With all these steps, we can elute the desired proteins. The viscosity can be different for each solution. The non-specific binding can be reduced by the addition of glycerol. Each protein purification process needs several steps to achieve pure proteins. The separation efficiency can be based on the processing time. It is important to alter the concentration of the affinity beads and protein concentrations. If we add lesser concentrations of affinity materials, there will be incomplete binding. The binding of other proteins can be possible if there are minimal affinity beads. The non-magnetic agarose beads can be refined from a medium by the process known as centrifugation or by filtration. The filtration process can be common in case of gravity columns. In the case of chromatographic systems, it is common. Changing these methods needs high levels of optimization. In the case of low levels of protein expression, we need to use higher concentrations of starting materials. In the case of magnetic beads, the separation principle can be independent on both, volumes and mode of application. The sensitive applications or test reactions like immunoprecipitation can be done just by using microliters. It is important to select an appropriate matrix to purify proteins. The selection of the purification process is important. The processes such as pull-down assays and immunoprecipitation can be important to enrich the protocol efficiency. The selection process must be based on several variables. This is mostly based on the properties of the biomolecules. The matrix selected must enhance the protein yield and the accessibility should be high. We can use paramagnetic beads in this separation process. Magnetic Sepharose beads are made up of highly cross-linked agarose particles. These beads comprise of the surface ligand. This will bind with the target proteins in the samples. This can be direct or indirect and it is based on the selection method.

There are many advantages ofthese magnetic beads. These beads can be non-adhesive and we need not add any detergents for this purpose. The collection process will be simple as there will be high visibility and density. Most of the beads are collected with magnetic devices. The centrifugation process may not be needed. The overall collection time can be drastically reduced. In some cases, it takes just seconds to isolate desired proteins. The target proteins may be concentrated from the ml range to microlitre scale. There are much functionalities that can render the usage of these beads useful and cost effective. These beads can be easily used in the purification process. Mostly, these beads can be classified into beads with immobilized ligands and pre-activated beads that bind with the ligands. The mode of selection can be based on the downstream applications, target proteins and assay to be used. In this case, the pre-activated beads can be used as they are flexible. The binding of ligands will be at definite sites. The beads having specific groups on the surface can be employed in case of pull-down assays. We need to use a ligand that can be linked covalently with the matrix. The beads can help in purifying the target biomolecules via affinity. The target will be eluted but the ligands can still be attached with the beads. The beads that are with the immobilized ligands can be easy to use. Most of the antibodies can be linked with the surface of the ligand. The antibodies can be eluted and so the desired molecule. We may not detect the molecules that are in trace amounts. The molecules that are used in the signaling pathways may not be detected by MS or electrophoresis. We can use Sepharose beads so that the targets can be captured. The buffers need to be optimized so that the binding efficiency will be enhanced. The buffering parameters can be based on the composition of the desired proteins. It is an important task to select the groups that are on the surface. The protocols must be modified accordingly. The magnetic beads can offer many advantages in life science research.

The microsphere technologies can use both magnetic and non-magnetic spheres. These spheres can be semi-porous or solid. In some cases, we can use cracked microspheres. The binding capacity of the desired proteins can be limited. The microsphere technology constitutes a hyper porous matrix. This matrix allow the penetration of the synthetic and biological molecules on the beads. In the conventional bead techniques, there will be limited binding seen on the surface. The solid bead can offer a low surface area. There will be a limited binding capacity. The performance will be reduced. The porous or cracked bead can enhance the surface area. The surface area will be improved and it is better than the solid beads. There is a novel sphere type named Resyn bead that can offer a polymer network. There will be an unparallel surface area. The capacity is higher than the other two. The magnetic particles in the purification of the proteins can offer us several benefits than the traditional techniques. The process is quite simple and the time to be consumed will be less. Several variables and factors can affect the process in many ways. The average size of the particles in the magnetic beads can be important in the purification process. If the mean particle size is less, then the surface area needed for protein binding will be high. Once the particle size is increased, there will be increased magnetic forces and the overall process will be diminished. With this knowledge, we can say that the processing time and the binding capacity are related to each other. As the users demand diverse needs, most of the suppliers provide diverse particle sizes. The distribution of the particle size can be important. The characterization of an image can be done with scanning electron microscopy and transmission electron microscopy. The aggregates and aggregated particles during the process of sample drying may not be visualized with these techniques. The disc centrifuge photosedimentation can be used as an important tool in knowing the distribution of the particles. The particles of various sizes can be permitted to approach the photodetector at different time spans. This can occur due to centrifugal forces.

Multifunctional nanoparticles are being used in the purification of the proteins. Nanoparticles can have diverse functions and they are used in many assays in life science applications and drug delivery systems. We can use magnetic nanoparticles to distribute the energy that can be oscillating.

Nanometer glues can be considered as an advanced technology in the protein purification protocols. The metal polymers can coat the particles like Qdots and magnetite. The Qdots can be about 10nm and the magnetite particles will be about 15nm. The particles can be coated with the polymers and the aggregation of the particles can be prevented. This can be due to charge repulsion technology. The charge repulsion technology can stop the formation of clumps in the particles that are activated. There are many benefits with metal polymers. These metal polymers can form a thin coating. This coating will be based on the metal type. The metal polymer can be cationic in nature. It can form a glue layer and it is thin. Mostly we can expect one metal ion thickness. Technically, the thickness will be about 0.2nm. The surfaces that are activated by the polymers can offer high rates of reproducibility. This can be due to the firm uniformity between two particles. There will be rapid kinetics existing between these particles.

7.2 FUSION PROTEIN PURIFICATION METHODS

The field of genetic engineering has offered us several opportunities to change the protein conformations and properties thereof. This demands an improvement of the protein purification techniques. Just by inserting DNA sequences, the amino acids sequences may change at the protein end. This can be altered to produce purification fusion. We can utilise these fusions to stabilize the proteins so that the proteases may not act on the proteins. The proteins can produce an inclusion body so that they can get some protection. The fusion can offer certain purification properties. This fusion can help us purify the proteins using ion exchange, metal chelate and immunoaffinity, partition and hydrophobic chromatography. The amino acids can be altered in the protein so that the patches will be formed. This will result in the development of affinity for a matrix. These developments in the process scale technology have helped several researchers in purifying proteins. The analytical methods may be needed in the digestion of protein fusion. Some methods can be needed to eliminate the product obtained due to digestion. Each protein has unique properties and designing the protein fusion may require certain inputs. These inputs are important in protein engineering and genetic engineering. APIII can be purified and the protein can be released using V8 protease.

Atriopeptide III can be important in the vasodilation process. This peptide can be used as a diuretic. This can be used in the treatment of hypertension. This peptide has 24 amino acids. This peptide cannot be stable in the *E.coli* systems. The instability of this peptide is due to protease activity. This peptide can be fused with RecA protein of bacteria so that it becomes stable. APIII doesn't have glutamic acid but it has aspartic acid. A gene was constructed and it expressed aspartic acid containing proteins at the junction of APIII and RecA protein. Due to this, the needed cleavage was obtained. Cost of the goods needed in the protein purification process can be a big issue. The food, protein polymer materials and biocatalysts can be produced in large quantities to satisfy the industrial and domestic needs. His-Tag purification system uses the affinity reactions of the histidine residues towards metal cations. We can use chromatographic support that can be prepared by a chelating agent with Nickle and Zinc ions. It can be used in IMAC or immobilized metal affinity chromatography (Yip et al., 1992). The fusion of polyhistidine can exhibit higher affinity to these supports. The fusion proteins are purified in a single step. The polyHis-tagged proteins can get associated with the IMAC systems. This is possible only if the salt concentrations are high and so the denaturants. Due to enhanced ionic strength of the buffers, there will be an increase in the multi-site attachments (Wei Jiang et al., 2018). The magnetized particle separation technology can be used with IMAC. In this technology, IMAC will be linked with the magnetic agarose beads covalently. This linkage will be facilitated by iminodiacetic acid. There will be a reaction with the metal ions. This technique was used in isolating polyhistidine-T_4 lysozyme from extract with copper ions (O'Brien et al., 1981).

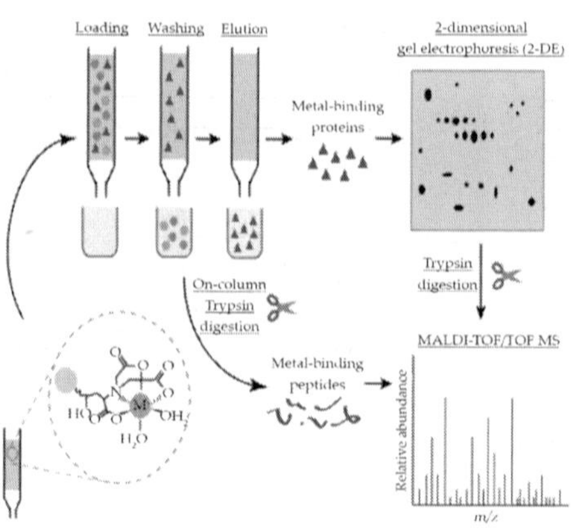

Figure 7.1. Immobilized metal affinity chromatography.

Source: https://www.sciencedirect.com/science/article/pii/B978012803814700 0095

The somatotropins can be classified as protein hormones and they are important in regulating the growth patterns in the animals. The human growth hormone or HGH can be important in treating dwarfism. It can be used to enhance lactation in the cows. The precipitation process can be employed to concentrate the dilute proteins that are impure. This can be done by regulating the precipitating agents. This can be used in the selective precipitation of the proteins and elimination of the impurities in the solution. This process can be important in the large-scale production of the proteins. We need a centrifuge and a stirred tank for this purpose. We can use this technique in the purification of soya proteins. When the crude samples are used, we can use the precipitation reactions of the fused proteins. We can use polyarginine proteins in this case (Ford et al., 2019). It is important to isolate proteins in a pure state so that the function of the proteins can be studied. The speed and convenience of the process is more important than the cost of the goods. It is important to engineer the proteins so that we can get the required properties. Sigma has produced a kit that can help in the purification of the proteins. In the purification process, we need to have a polyHistidine tag made of 6 histidines. This tag can exhibit a high rate of affinity towards divalent metal ions. There will be a chelate formation. The polyHistidine tag may not occupy the sites of the metal. These metal ions can be linked with the fusion proteins. This complex can be purified with IMAC. In this technique, NTA or Nitrilotriacetic acid is immobilized on agarose. polyHis will exhibit a high affinity towards nickel and so this process can be highly effective. Sigma is a kit developed jointly by Kodak and Immunex. In this system, the purification and expression processes can be facilitated. The recombinant proteins can be purified in yeasts and *E.coli*. FLAG peptide can get encoded at N or C terminus of the desired proteins. FLAG peptide can be classified as an octapeptide. This is designed to facilitate detection and purification. The FLAG expression kit includes enterokinase, control plasmid and protein, affinity gel FLAG peptide, anti-FLAG monoclonal antibodies, sequencing primers and expression vectors.

7.3 INCLUSION BODIES

The recombinant DNA technology helps us in the production of several proteins that are in limited proportions with the natural sources. rDNA technology has allowed us to produce proteins that are therapeutically valuable. These proteins can be produced on a large scale due to rDNA technology. The homologous and heterologous proteins can be expressed at high levels in the production of insoluble aggregates. These insoluble aggregates are called as inclusion bodies. These inclusion bodies are present in the cytoplasm. The recombinant proteins can be expressed in *E.coli*. The reason for choosing *E.coli* is that about 50% of the cell protein can be the expressed protein. We can expect more inclusion bodies from the cytoplasm. There may be more than one inclusion body in each cell. Production of proinsulin is the best example to explain the formation of inclusion bodies whilst the expression of heterologous proteins in *E.coli* (Williams et al., 1989). The accurate mechanisms that guide the production of the inclusion bodies are not yet clear. It was found that the inclusion bodies are also formed in other organisms. The inclusion bodies were found in insect cells, *Bacillus* sp. And *Saccharomyces* sp. (Thomas et al., 2010). The protein characteristics, media composition, growth temperature, protein expression and rate of expression can show effects on the production of the inclusion bodies. Just one parameter alone cannot affect the proteins.

The enzyme EcoRI can make clear the expression levels that are needed to produce the inclusion bodies. This enzyme will be present in the soluble proteins but in small amounts. If the expression level of this enzyme is high, we can expect the formation of the inclusion bodies. It was recorded that low temperatures can affect the protein concentrations in the soluble form in the cytoplasm and the same condition holds good for several proteins produced in *E.coli* (Schein, 2010). The inclusion bodies can be recovered from the cell homogenates and this process will be done by filtration and centrifugation. The inclusion bodies are dense and so we can use differential centrifugation of the cell debris. These homogenates can have many constituents. Some of them are insoluble cell debris, soluble proteins, inclusion bodies, nucleic acids and membranes. Most of the inclusion bodies can be recovered from the homogenates via large scale centrifuges. We can perform centrifugation at 20000 g for 30 minutes so that the sediment may have inclusion bodies. Most of the inclusion bodies can be extracted from the soluble materials. We can use a continuous disc-stack type of centrifuges to recover these inclusion bodies on a large scale (Wong et al., 1996). We can use the filtration technique to recover the inclusion bodies and the recovery rate is the same as the centrifugation process. It is important to wash the inclusion bodies so that the proteins will be refolded. Post washing, the proteins will be purified. Washing must be done to eliminate the protease enzymes so that the instability will be prevented in the purification phases. Most of the washing processes eliminate DNA, soluble cellular proteins and lipids. These processes must be done prior to the solubilization process. We can even reduce the pyrogens during the washing stage. There are many solutions in the market that are meant for washing to eliminate the contaminants present along with the inclusion bodies. Wash solutions such as deoxycholate, EDTA and lysozyme can eliminate the nucleic acids, phospholipids and lipopolysaccharides (Langley et al., 1989). The inclusion bodies were isolated for the first time via centrifugation post washing. Octyl glucoside, deoxycholate, urea, sucrose and Triton X-100 were used to eliminate the contaminants. We need to make sure that the pH is appropriate in the washing process. Some proteins are dominant in the inclusion bodies. They are OmpA, OmpC and OmpF that are proteins present on the outer membranes (Rinas et al., 2007). In these bodies, we can find 2 heat shock proteins or HSPs. These bodies harbor kanamycin phosphotransferase (Schoner et al., 1987), elongation factor Tu (Kane et al., 1992), large ribosomal subunit L13 and RNA polymerase subunits (Hartley et al., 1981). We need to use protease inhibitors in the buffer used for washing and so we can avoid proteolysis. Several factors can determine the washing efficiency. Some of the factors are additive concentrations, buffer to protein ratio, homogenization efficiency, exposure time, temperature, ionic strength, pH and stability of the proteins.

7.4 AUTOMATIC SMALL-SCALE PROTEIN PURIFICATION

The protein purifications at small scale can offer us immense opportunities in the process development. It is important to go for miniaturization in the case of protein purification as it will reduce the time and simultaneous processing of the samples is possible. This technique can enhance the statistical significance of several variables. The development of biotherapeutics comprises of expression and purification steps at small scales and there will be a downstream characterization of several biomolecules. Due to the scaling down of the processes, the design space will be more than enough. Though there are several advantages of scaling down, the relationships between the quality attributes and process parameters should be experimentally designed. Mostly, the process development projects use a design by experiment or DoE or quality by design or QbD. They may need the tendency to screen complex matrix conditions that are important in the identification, structure and titer of the protein functions (Swalley et al., 2009). If we need significant statistical analyses, we need to get enough sample numbers. The sample numbers can be obtained by the analysis and purification of the recovered molecules that can be classified as critical. Mostly, the miniaturization process in the analysis and purification systems can lower the processing times. Due to this, we can easily and simultaneously process several samples. This will lead to an enhanced statistical significance among the variables (Vincentelli et al., 2011). Most of the miniaturized protocols or procedures are predictive and this will help us in the scaling-up process. The lab-scale columns can be better than the minicolumns. In this process, the velocity-based mass transfer can be considered as the rate limiting step (Lacki et al., 2011). Most of the predictive capabilities of the columns that are miniaturized can be impacted by nonlinear elution gradients, wall effects, flow rates, residence times, the height of the bed columns and several geometric factors. The transport properties are managed by proteins that are greater than 150 kilodaltons (Welsh et al., 2011). We can enhance the throughput with this small-scale purification process. The bottleneck of this can be protein analysis and characterization. It is important to have inexpensive analytics to solve the sample number problem that needs to be addressed in QbD. The programmed methods can help us in the complete automation of these techniques. There was a huge development in the fermentation design, cell engineering and molecular biology in the last two decades (Shukla et al., 2010). Despite the several attempts made to get cost-effective protocols in protein purification, still, there l remain many challenges regarding the purity and nativity of the proteins (Singh et al., 2007). At the pilot scale, we can observe affinity domains like maltose binding protein or MBP. Chitin Binding Domain or CBD and poly-histidine that can be used in the purification methods (Kong et al., 2012). There are affinity resins that can provide us with high rates of purity. Most of the methods have limitations due to the high costs of the resins. The solution for this was the use of biopolymers. The biopolymers can be used to produce purification methods that are not considered as chromatographic techniques. The biopolymers that are engineered can be classified as purification tags. We can use these tagged fusion proteins so that the aggregates are highly selective in both physical and chemical conditions. Some of the tags are ELK16, repeat-in-toxin (RTX) domain and Elastin-like polypeptides or ELP (Banki et al., 2005). The ELP tag can help in the aggregation of the fusion proteins in high salt concentrations and high temperatures (Fong et al., 2009). Many lab techniques can be used to remove the tags. Some of the molecules that can be used are proteases. They are Rhimovvirus 3C proteases, TEV protease and Factor Xa (Raran-Kurussi et al., 2017). Inteins can be considered as cost-effective (Fong et al., 2009). The inteins are engineered so that there will be cleavage at high pH conditions.

7.5 NEXT-GENERATION CE

The capillary electrophoresis linked with mass spectroscopy can be useful in profiling certain compounds in metabolomics. The ultimate goal of the metabolomics is to solve clinical questions and biological complexities (Ramautar et al., 2016). It is important to use advanced analytical techniques so that there will be metabolic profiling of the samples (Kuehnbaum et al., 2013). The metabolic profiling of the metabolites can be done with MS and chromatographic techniques. When the availability of the samples is limited, metabolomics plays a vital role (Nevedomskaya et al., 2010). We can use CE-MS in the analysis of ionogenic metabolites especially, with the biological samples (Ramautar et al., 2016). With this technique, there will be less sensitivity and reproducibility. CE-MS can be used in the global profiling of endogenous metabolites and native peptides (Pontillo et al., 2016).

8

Protein Analysis

CONTENTS

8.1 INTRODUCTION

Ion mobility is a property that can be used as an analytical technique in the separation of the biomolecules and this was superseded by the progress made in proteomics (Valentine et al., 2006). Metabolomics has also produced several products that are far better than those produced from conventional techniques. There will be a separation of the gas-phase ions that is mostly dependent on the buffer gas mobility. The separation process can be linked with the size, shape and charge. The hyphenation of IMS or Ion Mobility Spectrometry can be called as IM-MS or Ion Mobility- Mass Spectroscopy. We can measure the ion mobility to know the averaged collision cross-section. This information can help us in finding the size and shape of molecules. We can use kinetic energy to derive a relationship between the collision cross-section and ion mobility (Mason and McDaniel 1988). The IM-MS can be effective in determining the structural analysis of the proteins and their assemblies (Snijder and Heck 2014). We can use the measurements of the collision cross-section and structure to analyze bulk molecules. The cross-sections that are measured can be loaded into a database. In this database, we can find sequence information and molecular weight of the peptide. We can use this data to know the cross-section of various amino acids. It was observed that the increasing molecular weights can be associated with the increasing cross-section. With this information, it is obvious that we can identify peptides. The size of the intrinsic amino acids was defined by the peptide ions that were doubly charged. This was observed in the yeast proteome. IMS can be used in both the separation and profiling of the proteins present in plasma (Liu et al., 2007).

8.2 PROTEIN THERAPEUTICS

The first recombinant protein is human insulin and it was produced in 1978. Protein therapeutics have involved in producing treatments for several diseases and disorder like cancer and other infections (Scott et al., 2012). There are almost 130 proteins that were approved by the FDA or Food and Drug Administration of the US (Leader et al., 2008). Most of the consumption is in the ADCs or Antibodies and antibody-drug conjugates (Voynov et al., 2009). Most of the protein-based and peptide-based therapeutics have emerged into clinical stages and there is always a demand for these drugs. These applications involve pharmacokinetic studies and the quantification of drugs. Bioanalysis data can be considered important in the development of drugs for particular ailments. The protein bioanalysis in the case of therapeutics needs assays. These assays must exhibit a high sensitivity and selectivity. It is important to ensure that there is good accuracy and precision in these assays. Through these assays, we can expect a high rate of throughput. Two important techniques can be used in the protein therapeutics especially in the bioanalysis. They are Liquid Chromatography-Mass spectrometry or LC-MS and Ligand Binding Assays or LBAs. The Liquid Chromatography-Mass Spectrometry can analyse the analyte concentration. This analysis can be done by studying interactions between the analyte and reagents. These LBAs can be used in the quantification of the proteins that are present in the biological matrices. We can use LBAs because they provide throughput, robustness and high sensitivity (Damen et al., 2009). There are some disadvantages reported lately. This technique is quite expensive and time-consuming. The LBAs that are nonspecific can quantify metabolites and drugs. Due to minimal selectivity, we may not demarcate between drug and metabolites. The accuracy can be reduced due to the presence of endogenous components. This may be due to antibody and antidrug formation or ADA formation (Zheng et al., 2014). The LC-HRMS strategy can help us in solving these problems. In the case of peptide and protein analysis, the liquid chromatography and tandem mass spectrometry can be used. This technique can provide us with several advantages over conventional techniques. We can demarcate the metabolites and isoforms. We can even identify missing amino acids. In the case of LBAs, it is highly impossible to demarcate these components. The tandem MS fragmentation technique can produce fragment ions in the spectrum. This spectrum can be used in the quantification process. The quantification of the analytes can be done with a high rate of precision via unique fragment ions (Wang et al., 2010). Liquid chromatography-selected reaction monitoring or LC-SRM can be considered as an important tool in the analysis of small molecules (Zhou et al., 2005). In a mass spectrometer with a triple quadrupole, there will be 3 quadrupoles. They are named as Q1, Q2 and Q3. Q1 can be used to detect the precursor ion. Q2 can represent the collision cell. The precursor ion can be fragmented into the ions at the collision cell. Q3 can serve as a mass filter. This permits the identification of the selected fragment ions that may be detected. Due to the monitoring of the transitions between fragment ions and precursor, the rate of selectivity can be enhanced. We can expect high sensitivity for analytes. This technique has some problems in case of bioanalysis of protein therapeutics (Ewles and Goodwin., 2011). Extensive modifications and metabolism can make the process difficult and the optimization of the instrument will be a big problem. High selectivity is needed as there will be interferences from the endogenous proteins. This is true with the biological materials. There will be high background noise and so the selectivity will be compromised. Because of this, we may not detect the peptide targets that are in trace amounts (Lange et al., 2008). We cannot quantify the intact proteins. This can be due to low fragmentation efficiency. Due to its low acquisition capabilities, large proteins like monoclonal antibodies may not be detected. These antibodies may have multiple charges and the charge distribution is wider. With this instrument, there will be 1500 acquisitions, and that may not be enough to investigate the peptides with high mass. Due to these limitations, LC-MS methods are being used to quantify the peptides, especially from the biological samples. The analytes will be subjected to enzymatic digestion.

Most of the mass spectrometers that can provide high resolutions may include TOF or time-of-flight, Orbitrap and FT-ICR or Fourier transform ion cyclotron resonance. The quantitative analysis can be limited with the conventional mass spectrometers. We can yield a high resolution, faster scan speeds, wide dynamic range and enhanced sensitivity with HRAM or high-resolution accurate mass orbitrap mass spectrometers. The HRMS systems have more advantages over SRM systems. In these systems, there is no need of specific tuning for a compound. With this latest technique, we can perform qualitative and quantitative analysis simultaneously (Korfmacher 2011). In this technique, the ions can be trapped in electric and magnetic fields of FT-ICR. The trajectories will be circular and mostly perpendicular with the magnetic fields. The ions that are excited can be in motion in a circular fashion. Their decay can be identified in the form of domain signals. These signals can be transformed into the mass/charge ratio (Marshall et al., 1998). The detection of the image current cannot lead to ion loss. The ions in these systems are stored in ICR for prolonged periods. This will result in greater mass accuracy and high resolution. This technique needs a strong magnetic field and a large space for storage. The optimization times are long and because of these factors, the maintenance can be expensive. Due to these limitations, we cannot use this technique for regular lab routines.

Time-Of-Flight is employed in these spectrometers. TOF has a huge dynamic range and we can expect faster scans. The separation of ions will be dependent on the velocities. The traveling time of the ions through a flight tube towards the detector may be determined. Most of the smaller ions can be detected first. We can increase the flight path so that the resolving power will be high. We can used reflectron TOF rather than linear TOF as the former yield high resolution. The sensitivity can be reduced or even lost in some cases. This is due to the extended travel time via flight tubes. There are four methods which can be used in the quantification. They are Data-independent Acquisition or DIA, full scan, parallel reaction monitoring or PRM and selected ion monitoring or SIM. The last two methods can be classified as targeted quantification methods. The Data Independent Acquisition method and full scan can be classified as quantitative and qualitative methods. DIA and full scan can be classified as non-targeted methods. Based on the sensitivity needed, protein size and drug development, there are 3 basic approaches in quantifying the proteins. They are bottom-up, intact and middle-up approaches.

In the case of bottom-up approach, the proteins can be quantified at the peptide levels. This technique can be classified as an important method. It can provide accurate quantification and high sensitivity. In this technique, the proteins will be subjected to proteolytic digestion. When monoclonal antibodies are used, denaturation will result in the unfolding of the proteins. The disulfide bonds will be reduced by DTT or dithiothreitol. The alkylation process will be done by IAA or iodoacetamide.

It is important to use proteases at the cleavage sites so that the purification process will be accurate. In the Bottom-up approach, the target analyte will be peptides. With this approach, we can expect high sensitivity, linear and wider dynamic range and accurate quantification. Problem with this technique is the introduction of extreme modifications. The acquisition methods used can be DIA, PRM and SIM. In Middle-up approach, the target analytes will be antibody fragments. They can be light chains and heavy chains or Fab/Fc. With this approach, we can expect to restore the original sequence information partially. Accurate quantification is possible with this technique. The sample preparation process is quite simple. The main limitation of this is a low dynamic range. The acquisition method can be a full scan.

In the intact method, the target analyte will be intact proteins. With this approach, we can expect spectral information and mass data. There will be less modifications post the enzymatic digestion. The sample preparation is simple. We get a low dynamic range that is linear. Due to the wider distribution of charge, we can expect low sensitivity. Mostly, the accuracy can be affected and this is due to truncation and modifications. The acquisition technique can be a full scan. The quantification rate with LC-MS can be affected by several factors. The improvement of the sample preparation can enhance the sensitivity of the LS-MS. The signal suppression and background noise can be due to albumin, immunoglobulins and endogenous proteins. The matrix can affect the sensitivity of LS-MS assays. We can use protein precipitation to eliminate contaminants in the serum and plasma (Anderson and Anderson 2002).

8.3 ORBITRAP MASS SPECTROMETRY

The protein therapeutics was in the market in the late 1920s. Insulin was considered as the first therapeutic protein produced. In those days, insulin was purified from the pancreas of animals. Insulin has saved many diabetics. Due to animal-derived insulin, there were several issues produced. Some of the issues were immunogenicity, cost and availability. Recombinant proteins are now the protein therapeutics and they use recombinant DNA technology. Since the 1980s, people started using recombinant human insulin and it was a big success (Pavlou and Reichert 2004). The recombinant monoclonal antibodies were classified as effective drug candidates and are human therapeutics. They have many advantages. They can stimulate immune cell effector responses, low toxicity levels, high selectivity and lifetime circulation. Many techniques can be used to investigate the minor variants and protein modifications. Electrophoresis and chromatography can be considered as important methods that can be used in this investigation. The proteins can be characterized by MS at intact protein levels and even at peptide levels. Tandem spectrometry can be used to disperse the intact proteins into fragment ions. This technique can be used in detecting site-specific modification of proteins. It will be a hard task if the size of the proteins exceeds 30 kDa (McLafferty et al. 2007).

Mostly, a MS instrument comprises of data processing electronics, detector, a mass analyzer, optics and an ion source. The ion trap analyzer can separate ions based on their oscillation frequency (m/z). The quadrupoles can be used to separate the ions and it is based on the stability. Orbitrap MS can be classified as an important tool in the identification and characterization of the proteins. In the MS/MS techniques, the precursor ions of peptides will be selected. Then there will be a production of ions that are to be detected. Oxidation can produce protein modification and it is non-enzymatic in nature. This can occur whilst processing and storage. Cysteine and Methionine can be sensitive to oxidation. In the case of therapeutic monoclonal antibodies, the oxidation of Trp and Met can be important. These modifications are common in the formulation and storage phases (Ji et al., 2009). Deamidation can be another nonenzymatic protein modification. Mostly asparagine

(Asp) can be deamidated. These modifications can impact both structure and function of the proteins. It is important to know the level at which the deamidation occurred. At basic pH conditions, we can expect enhanced deamidation levels (Song et al., 2001).

8.4 QUANTITATION OF PROTEINS

The biotherapeutics can be classified as an important drug class that is making progress drastically. On the evaluation of the LC-MS bioanalysis, there are certain problems associated with the accuracy and precision of the quantitation. The paramount importance must be given whilst sample preparation. Some of the challenges in the bioanalytical quantification are limited quantitation range, reduced sensitivity, low specificity, coelution, diminished data quality, low sensitivity and recovery, peptide physico-chemical properties, reduced MRM selectivity, measurement errors and poor fragmentation of peptides. The cyclic fragments may not be subjected to complete fragmentation. Due to this, there will be a production of ions whilst analysis. In the case of low-level quantitation, we can expect some measurement errors. This will produce problems with the precision and accuracy of data. We may not obtain a satisfactory signal to noise ratio and so the MRM selectivity will be limited. The complex state of charge, low solubility and nonspecific binding can be considered as important. The polarity on the peptides and adsorptive properties can affect the recovery rates and interfere with most of the biological matrices. The selectivity and sensitivity are severely impacted. The accuracy and precision can be important whilst dealing with the low peptide levels. Most of them will not yield appropriate bioanalytical standards. The interference caused due to multiple charges in the coelution can diminish the accuracy of quantification. The isobaric interferences can reduce the specificity and selectivity of the assays. Due to complexity in the biological matrices, there will be low specificity and so the data resolution will be reduced. Due to complex matrices, the sensitivity will be impaired. In the case of limited peptides, the background can suppress the actual sensitivity needed. The quantitation range will be limited. ELISA may not be satisfactory in providing accurate bioanalysis. Due to this LLOQ or lower limits of quantitation will be compromised. Optimization in the sample preparation steps can enhance the quantitation process. The biopharmaceuticals can be classified as important therapies recommended.

8.5 INTEGRATED CST

Several biomarkers can be used in the treatment of several diseases. The biomarkers comprise of metabolites, peptides, proteins and genes. Due to their diversity, the analysis of biomarkers is often challenging (Heger and Kellogg 2014). Mass spectrometry is considered as an important tool in the development of qualitative proteomics. The nanobore liquid chromatography or nLC and Electronspray ionization mass spectrometry or ESI-MS were used in the analysis of peptides and proteins of the biological fluids, tissues and cells (Washburn et al., 2001). The biomarker discovery and Nanobore based LC-ESI-MS platforms in investigating proteomics can be identical. In LC-MS, there is reduced column diameter, minute particle sizes and low flow rates (Lee and Kerns 1999). The combination of nano-LC and micro-LC equipped with ionization of nanospray can be useful in bioanalysis with high rates of sensitivity.

8.6 TARGETING OF ISOFORMS

Most of the human genes can be transcribed into pre-mRNAs or precursor-messenger ribonucleic acids. Most of the pre-mRNAs are subjected to post-transcriptional modifications. Mature mRNAs can be produced via alternative splicing. In this process, the introns will be removed and the exons will be sealed. A single gene can produce multiple forms of mRNA. These mRNAs can produce splice variants with diverse amino acid sequences (Tavers et al., 2014). Due to alternative splicing, there will be an increase in the diversity of messenger RNAs. This is the reason behind diversity in the proteins being produced in humans. Most of the protein functions will be affected by alternative splicing mechanisms. It is important to know the structure, function and expression of the isoforms produced due to the alternative splicing process. This knowledge is essential in the development of biomarkers and therapeutics. The functional studies and expression patterns of the splice variants in the disease conditions can be useful (Lisitsa et al., 2014). The pre-mRNAs can produce proteins with altered structures and sequences. The alternative splicing mechanisms can regulate the metabolic pathways, lipid metabolism, coagulation of blood and development of the heart. Several therapeutic drugs can comprise of many splice variants. Latest knowledge of the splicing mechanisms is mostly from the RNA transcripts (Omenn 2014). There are many approaches that can investigate the expression profiles and splice variants in diseased and normal states. The profiling of the splice variants can give us more information about the signaling pathways and disease mechanisms. Mass spectroscopy has helped us to quantify the proteins via proteolytic digestion (Yassine et al., 2013). The stable isotopes can be used in multiple reaction monitoring. The affinity capture can be linked with MRM-MS and this has become an essential tool for quantitation of the therapeutic targets and endogenous biomarkers. This can be done by validating the proteolytic peptides. OPN can be classified as a glycoprotein that is secreted whilst metastasis of various cancers (Gimba and Tilli 2013). If the OPN levels are higher, then we can say that the progression of the disease is high. The survival rates can be decreased then. This can be classified as a therapeutic target and a diagnostic biomarker. The cancer specific isoforms of OPN was generated by immunoaffinity-MRM-MS method; this method is being used to quantify OPNa-c isoforms in both diseased and healthy persons (Wu et al., 2012). In this method, the antibodies recognize OPN sequences and so most of the biases will be eliminated. Trypsin can be immobilized and it helps in the effective digestion of the peptides. We can use stable isotope-labelled parts in determining the isoforms. We can use phosphatases to restore the peptides that are at native state. The splice junction peptides can be detected by nano-LC-MRM-MS.

8.7 BIOTHERAPEUTICS AND PROTEIN BIOMARKERS

The proteins can help us in performing several biological functions. It is important to measure the proteins in order to understand the essential physiology of the biological systems. The protein biomarkers can be important in the translation of the therapeutic agents clinically. The biomarkers can be used in determining the pharmacodynamic endpoints. This is important in dose selection and target engagement. We can use biomarkers in deciphering the disease progression in patients. Mass spectrometry can be used in the analysis of complicated proteins. Multianalyte analysis can be performed by coupling chromatography with Mass spectrometry. We can get both quantitative and qualitative data with these techniques. The measurement of proteins in the specimens can be through LBAs or ligand binding assays. In the case of solid tissue samples, it is important to use immunohistochemical staining. IA-MS methods can be divided into 2 groups. IA-MS methods can be classified based on application requirements. An intact peptide can be a target for IA capture and it can be used in detecting the biomarker peptide. There will be no enzymatic digestion and so we call it as top-down analysis.

Several LC formats can be used in investigating the proteins. There are several variants in LC formats. Multidimensional configurations, flow rates, column diameters and stationary phases get varied with LC's.

There are several applications of IA-MS. The peptide biomarkers can be <10kDa. They are involved in almost all the biological processes. These biomarkers can help us in the identification of the disease states in humans. The protein targets will be >10kDa. The sensitivity with MS can be an issue and there will be bottom-up analysis. The protein biomarker quantification can be done with anti-peptide IA-MS. It can be classified as a complementary strategy. This technique was devised by Anderson (Anderson et al., 2004). SISCAPA is an important label used in IA-MS applications. It is associated with bottom-up proteomics. The translation of biomarkers in most of the clinical applications is called as verification. The term 'verification' was coined by Rifai et al (Rifai., 2006). The polypeptides and peptides can be used in many biochemical processes that can serve as antagonists and agonists. Antibody-drug conjugates can be classified as the latest class of biotherapeutics. They are flexible. The ADC can be subjected to internalization in the cells. The problem with these ADCs is that they can get easily released in the systemic circulation.

8.8 IMMUNOCAPTURE LC/MS

Immunogenicity represents the immune responses generated by both animals and humans against antigens. Multiple factors can be involved in the production of immunogenicity against the biotherapeutics. Certain factors can be specific to the patients. In case, biotherapeutics inflict immune responses, it will be a problem. The ADAs or antidrug antibodies can be produced by the host's immune system and this will alter the pharmacokinetic profile of the drug. Because of this, there will be an elimination of the therapeutic proteins in the blood circulation. We can use bridging LBAs in the assays. It is important to go for multiple assays in the assessment of immunogenicity by LBAs. This process requires more time and it is a hard task. The sensitivity and performance of an assay can be based on the quality and availability of the reagents (Butterfield et al., 2010). The pharmaceutical industries have shifted their main scope of business from the small molecules to the therapeutics, mainly protein therapeutics. This change is evident since last 20 years. The main reason behind it is that these therapeutics exhibits a low toxicity and high specificity. Some events occur in the lifetime of the therapeutic proteins. They are fragmentation, isomerization, deamidation and oxidation. The characterization of the isoforms can be considered as an important property with these protein therapeutics. We cannot categorize the individual protein variants as it is not easy to isolate variants individually. The properties of the protein therapeutics can be impacted by aggregation, denaturation and several conformational changes. We can get products with a high rate of purity only though flawless manufacturing procedures. The purified therapeutics alone can be used in clinical trials. The formation of the disulfide bonds can be important so that there will be stability in the protein conformation (Betz 1993). There will be the formation of disulfide bonds due to the shuffling of bonds (Hwa et al., 1999). The formation of the disulfide bonds can be linked with the pathways that control protein folding (Mamathambika and Bardwell 2008). The disulfide bond measures about 2 angstroms. The bond distance can be used in the prediction of the disulfide bonds in proteins especially within their 3D structures. In the case structures are not available at hand; we take the distance parameter as 8 angstroms between alpha carbons of two Cys residues. This can be used to define the potential disulfide bonds. The accuracy that is expected with this method can be closer to 80% :that is considered good (Mallick et al., 2002). We can use mass spectroscopy and Edman sequencing in the detection of the peptides linked with disulfides. We can determine the connectivity of the cysteine residues via enzymatic digestion. If the cysteine residues are too close then we may not be able to resolve the proteolytic fragments. In this case, it is important to use chemical modifications that involve sulfhydryls (Gray 1993). Due to the presence of the disulfide bonds, there will be reduced sequence coverage by mass spectrometry. The fragments produced due to cleavage between the Cys residues can be attached by the disulfide bond formations.

An absence of fragmentation between the cysteines and mass shift can be used as an important parameter in determining the disulfide bond formation. This can be done by electron capture dissociation method of MS (Han et al., 2006). The conventional method of analyzing disulfide bonds can be via proteolytic fragments. We can achieve a signature isotope pattern due to the proteolytic digestion of isotopic oxygen (Gorman et al., 2002). The disulfide isoforms can be demarcated via the connected methods of cysteine residues. In a protein that has 4 cysteines, there is a possibility of 2 sulfide bonds. We can expect at least 3 isoforms. With the increase in the disulfide bonds, we can expect an exponential increase in the isoforms. Most of these isoforms may not be isolated individually and this is due to their structural similarity. Several software packages are available in the market that can help us in assigning disulfide linkages by MS data (Murad and Singh 2013). In IgG isotypes, the disulfide structures are almost similar. We can expect a high sequence homology in these isoforms. In IgG2, the hinge region consists of 4 cysteines and they are adjacent to each other. There are about 3 disulfide isoforms that can be determined by MS and N' sequencing (Zhang et al., 2008). Due to their structural difference and conformations, these isoforms can be separated easily by ion mobility MS, electrophoresis and chromatography (Bagal et al., 2010). Oxidation can be classified as an important chemical modification that can occur in the protein therapeutics. This process can occur in the storage and manufacturing phase (Torosantucci et al., 2013). The oxidation process can be due to stressors. Common stressors are oxidizing agents, pH, light and electroactive metals. Theoretically, tandem MS can be used to obtain the sequence of glycan and peptides. We can even know the glycosylation sites with tandem MS.

The protein therapeutics can be considered as a fast-growing field over the last decade (Leader et al., 2008). Most of the translational research has converted into drug development research: the needs to test the transporters, enzymes, drugs and biomarkers. We find that there are 3 important strategies for the bioanalysis of proteins using LC-MS. They are surrogate peptide approach, intact protein approach and middle up or limited digestion approach. In the surrogate peptide approach, there will be a predigestion phase in the sample preparation step. Then the samples are subjected to digestion. This will result in the production of peptides. Then post digestion phase commences and the products generated will be allowed to pass into the LC-MS equipment. In the intact protein approach, the sample will be prepared and send to LC-MS directly. In the middle-up approach, the sample will be prepared and there will be limited digestion with IdeS protease. There will be a production of large fragments that are sent to LC-MS. The protein precipitation or PPT technique can be simple and efficient. This technique can be cheap. The samples can be cleaned easily. We can easily automate the process and this process is common whilst investigating the bioanalysis of drugs. PPT utilizes organic solvents that are miscible in water. Some of the solvents that can be used are methanol and acetonitrile. Large proteins such as monoclonal antibodies can get coprecipitated with other proteins from serum or plasma. We can expect trace amounts of analyte as the supernatant. The supernatant post PPT is not of used whilst dealing with large proteins. It is important to subject the pellet to digestion so that we can easily analyze easily. We can eliminate phospholipids, salts and soluble proteins with this technique (Ouyang et al., 2012). The solid phase extraction or SPE can be used prior to protein digestion. In the case of large proteins, we can use SPE (Heudi et al., 2008). We can further purify the samples using strong-cation exchange and 2D SPE. We can use orthogonal SPE for sample preparation. We can go for the derivatization process of the peptides in proteomics (Suh et al., 2010). Depletion of proteins can be used in the proteomics field and this will lower the complexities drastically (Whiteaker et al., 2007). This depletion is possible with immunoaffinity and chemical affinity. Protein digestion can be considered as an important step in the preparation of samples. This will ensure sensitivity, accuracy and reproducibility.

The cytotoxic molecules can be used in cancer treatment. We can expect poor therapeutic indices and toxicity to health issues. Most of the monoclonal antibodies can be target specific and are considered as candidates in cancer treatment. The antibody drug conjugates or ADCs can provide us a synergistic advantage over other techniques. ADC are made of a cytotoxin, chemical linker and mAb (Han and Zhao 2014). The structure of ADC comprise of antibody, cytotoxic agent and linker. The antibody portion of the ADC can exhibit reduced nonspecific binding. The antibody can bind to antigen that is expressed on the target cells. Antibody part of the ADC can exhibit high binding affinity towards antigens. The cytotoxic part of ADC can be highly potent if freed from the ADC. This part is non-toxic in natures when linked with monoclonal antibodies. This agent can be classified as non-immunogenic. We can expect a high rate of solubility in the plasma. The linker part of the ADCs can maintain toxic properties on reaching the cells that are to be destroyed. This part cannot alter the profile of the monoclonal antibodies. It can be stable even in the systemic circulation systems. ADCs that can travel in the plasma will be in the form of a complete molecule. ADCs can be in the circulation and it will bind to the cell receptor. Once ADCs reach the receptors, they will get internalized in the cell. There will be a formation of endosomes. These endosomes with ADCs can fuse with lysosomes. The linker present in the ADC is cleaved. The cleaved toxin will be released into the cell leading to cell death. ELISA technique can provide high sensitivity and we can detect picograms in ml. We can expect high throughput in ELISA. We can detect total and free antibodies. The main problems with ELISA are that this technique is based on the reagents and there will be matrix interference. With LC-MS and immunocapture techniques, we can expect high sensitivity and assay specificity. The development time is short and we can get a wide curve range. Some of the disadvantages are excess dependence on the reagents, limited validation and expensive equipment. In case of direct LC-MS with coprecipitation, we can expect multitasking, shorter development times, limited utilization of reagents and wide curve range. The assay specificity is high. Some of the problems are limited validation, total antibody detection, low sensitivity and expensive equipment.

References

1. Abderhalden, E., Defren, G., & Hall, W. T. (1908).Text-book of physiological chemistry in thirty lectures, by Emil Abderhalden, tr. by William T. Hall and George Defren. doi:10.5962/bhl.title.22148
2. Acikara, B. (2013, April 10). Ion-Exchange Chromatography and Its Applications. Retrieved May 20, 2020, from https://www.intechopen.com/books/column-chromatography/ion-exchange-chromatography-and-its-applications
3. Agarose gel diffusion (AGD; Ouchterlony test, Ouchterlony technique).(2015, 12).*The Dictionary of Genomics, Transcriptomics and Proteomics,* 1-1. doi:10.1002/9783527678679.dg00202
4. Alarie, J. P., Jacobson, S. C., & Ramsey, J. M. (2001, 01).Electrophoretic injection bias in a microchip valving scheme.*Electrophoresis,22*(2), 312-317. doi:10.1002/1522-2683(200101)22:23.0.co;2-3
5. Alnemri, E. S. (1997, 01). Mammalian cell death proteases: A family of highly conserved aspartate specific cysteine proteases. *Journal of Cellular Biochemistry,64*(1), 33-42. doi:10.1002/(sici)1097-4644(199701)64:13.0.co;2-0
6. Alton. (2014, July 15). Gel Filtration Chromatography. Retrieved May 20, 2020, from https://www.slideserve.com/alton/gel-filtration-chromatography
7. Amon, U., Amon, S., & Gibbs, B. F. (2000, 04).In vitro studies with fexofenadine, a new nonsedating histamine H1 receptor antagonist, on isolated human basophils.*Inflammation Research,49*(S1), 13-14. doi:10.1007/pl00000160

8. Analytical Chemistry. (2020, April 23). Retrieved May 20, 2020, from https://chem.libretexts.org/Bookshelves/Analytical_Chemistry
9. Analytical Ultracentrifugation.(n.d.). Retrieved May 20, 2020, from https://www.sciencedirect.com/topics/chemistry/analytical-ultracentrifugation
10. Andrews, G. F., & Fonta, J. (1985).The engineering design of continuous electrophoresis.*Electrophoresis,6*(10), 483-488. doi:10.1002/elps.1150061003
11. Anderson, N. L., & Anderson, N. G. (2002, 10).The Human Plasma Proteome.*Molecular & Cellular Proteomics,1*(11), 845-867. doi:10.1074/mcp.r200007-mcp200
12. Anderson, N. L., Anderson, N. G., Haines, L. R., Hardie, D. B., Olafson, R. W., & Pearson, T. W. (2004, 04). Mass Spectrometric Quantitation of Peptides and Proteins Using Stable Isotope Standards and Capture by Anti-Peptide Antibodies (SISCAPA). *Journal of Proteome Research,3*(2), 235-244. doi:10.1021/pr034086h
13. Aryal, S., & Shiferaw, W. (2020, January 02). Mass Spectrometry (MS)- Principle, Working, Instrumentation, Steps, Applications: Instrumentation. Retrieved May 20, 2020, from https://microbenotes.com/mass-spectrometry-ms-principle-working-instrumentation-steps-applications/
14. Aryal, S. (2020, January 05). Ouchterlony Double Immunodiffusion technique: Immunology. Retrieved May 20, 2020, from https://microbenotes.com/ouchterlony-double-immunodiffusion-technique/
15. Aryal, S. (2020, January 14). Polyacrylamide Gel Electrophoresis (PAGE): Instrumentation. Retrieved May 20, 2020, from https://microbenotes.com/polyacrylamide-gel-electrophoresis-page/
16. Autoradiography.(n.d.). Retrieved May 20, 2020, from https://lifeofplant.blogspot.com/2011/12/autoradiography.html
17. Baek, Y., Singh, N., Arunkumar, A., Borwankar, A., & Zydney, A. L. (2019, May 20). Mass Balance Model with Donnan Equilibrium Accurately Describes Unusual pH and Excipient Profiles during Diafiltration of Monoclonal Antibodies. Retrieved May 20, 2020, from https://onlinelibrary.wiley.com/doi/abs/10.1002/biot.201800517
18. Bakhtiar, R., Majumdar, T. K., & Tse, F. L. (2006, 05).The Role of Liquid Chromatography-Mass Spectrometry in Pharmacokinetics and Drug Metabolism.*HPLC for Pharmaceutical Scientists,* 605-640.

doi:10.1002/9780470087954.ch13

19. Ballantine, J. A., Games, D. E., & Slater, P. S. (1997, 04).The Use of Diethylamine to Determine the Number of Sulphonate Groups Present Within Polysulphonated Alkali Metal Salts by Electrospray Mass Spectrometry.*Rapid Communications in Mass Spectrometry,11*(6), 630-637. doi:10.1002/(sici)1097-0231(199704)11:63.0.co;2-k
20. Bandilla, D., Cabral, J., & Skinner, C. D. (2006, 08).Characterization of UV-transparent capillaries for CEC and CE.*Electrophoresis,27*(16), 3271-3276. doi:10.1002/elps.200600060
21. Banki, M. R., Feng, L., & Wood, D. W. (2005, 08). Simple bioseparations using self-cleaving elastin-like polypeptide tags. *Nature Methods,2*(9), 659-662. doi:10.1038/nmeth787
22. Banks, C. A., Kong, S. E., & Washburn, M. P. (2012, 12).Affinity purification of protein complexes for analysis by multidimensional protein identification technology.*Protein Expression and Purification,86*(2), 105-119. doi:10.1016/j.pep.2012.09.007
23. Barai, A. (2013, October 23). Chromatofocusing. Retrieved May 20, 2020, from https://www.slideshare.net/amlano/chromatofocusing
24. Bardelmeijer, H. A., Waterval, J. C., Lingeman, H., Hof, R. V., Bult, A., & Underberg, W. J. (1997). Pre-, on- and post-column derivatization in capillary electrophoresis.*Electrophoresis,18*(12-13), 2214-2227. doi:10.1002/elps.1150181212
25. Bartlett, S., Lagercrantz, H., & Smith, A. (1976, 08). Gel electrophoresis of soluble and insoluble proteins of noradrenergic vesicles from ox splenic nerve: A comparison with proteins of adrenal chromaffin granules. *Neuroscience,1*(4). doi:10.1016/0306-4522(76)90061-0
26. Belder, D., Deege, A., Kohler, F., & Ludwig, M. (2002, 10).Poly(vinyl alcohol)-coated microfluidic devices for high-performance microchip electrophoresis.*Electrophoresis,23*(20), 3567-3573. doi:10.1002/1522-2683(200210)23:203.0.co;2-3
27. Berger, T. A. (2015, 11).Instrumentation for analytical scale supercritical fluid chromatography.*Journal of Chromatography A,1421*, 171-183. doi:10.1016/j.chroma.2015.07.062
28. Berry, J. D., Yang, C., Fisher, J., Mendoza, E., Young, S., & Stupack, D. (2013, 12).Fc-Fusion Protein Expression Technology.*Therapeutic Fc-Fusion Proteins,* 45-66. doi:10.1002/9783527675272.ch02
29. Bhardwaj, S., & Day, R. (1997). Detection of Intra-cellular protein-

protein interactions: Penicillin interactive Proteins and Morphogene Proteins. *Techniques in Protein Chemistry,* 469-480. doi:10.1016/s1080-8914(97)80047-6

30. Bianchi, P., Macnamara, E., Bergami, M. R., Gasparro, C., Jones, R., Whicher, J. T., . . . Aguzzi, F. (1992, 02). Immunochemical Evaluation of Monoclonal Gammopathies: Heavy Chain to Light Chain Ratio is of Little Practical Value for Detecting IgD Myelomas and Free Light Chains. *Clinical Chemistry,38*(2), 317-319. doi:10.1093/clinchem/38.2.317a
31. Bio-Resource. (1970, January 01). Ion Exchange Chromatography Principle. Retrieved May 20, 2020, from https://technologyinscience.blogspot.com/2011/09/ion-exchange-chromatography-principle.html
32. Blake, M., Johnston, K., Russell-Jones, G., & Gotschlich, E. (1984, 01). A rapid, sensitive method for detection of alkaline phosphatase-conjugated anti-antibody on Western blots. *Analytical Biochemistry,136*(1), 175-179. doi:10.1016/0003-2697(84)90320-8
33. Bonanni, A., Pumera, M., & Miyahara, Y. (2011). Influence of gold nanoparticle size (2–50 nm) upon its electrochemical behavior: An electrochemical impedance spectroscopic and voltammetric study. *Physical Chemistry Chemical Physics, 13*(11), 4980. doi:10.1039/c0cp01209b
34.
35. Bílková, Z., Mazurová, J., Churáček, J., Horák, D., & Turková, J. (1999, 08). Oriented immobilization of chymotrypsin by use of suitable antibodies coupled to a nonporous solid support. *Journal of Chromatography A,852*(1), 141-149. doi:10.1016/s0021-9673(99)00234-4
36. Brumbach, J. H., Lee, Y. W., Kim, S. W., & Yockman, J. W. (2012, 04). Functional properties and biodistribution of poly(triethylenetetramine/cystamine bisacrylamide) and poly(triethylenetetramine/cystamine bisacrylamide)- poly(ethylene glycol) mixtures formed with nucleic acid. *Journal of Controlled Release,159*(1), 111-119. doi:10.1016/j.jconrel.2012.01.010
37. Büttgenbach, S., & Feldmann, M. (2008, 01).Application of UV depth lithography in micro system technology.*Optoelectronics Letters,4*(1), 1-4. doi:10.1007/s11801-008-7099-2
38. Caglio, S., & Righetti, P. G. (1993).On the pH dependence of

polymerization efficiency, as investigated by capillary zone electrophoresis.*Electrophoresis,14*(1), 554-558. doi:10.1002/elps.1150140184

39. Caldwell, J. C., & Caldwell, P. (1996).Toward an Epidemiological Model of AIDS in Sub-Saharan Africa.*Social Science History,20*(4), 559. doi:10.2307/1171342
40. Cameron, M. I., Morisco, D., Hofstetter, D., Uman, E., Wilkinson, J., Kennedy, Z. C., . . . Foster, J. M. (2020, January 22). Systematically Improving Espresso: Insights from Mathematical Modeling and Experiment. Retrieved May 21, 2020, from https://www.sciencedirect.com/science/article/pii/S2590238519304102
41. Carr, S. M. (n.d.). Retrieved May 20, 2020, from https://www.mun.ca/biology/scarr/Gr12-18.html
42. Carr, S. M. (n.d.). Retrieved May 20, 2020, from https://www.mun.ca/biology/scarr/iGen3_06-04.html
43. Castagnola, M., Rossetti, D. V., Cassiano, L., Misiti, F., Pennacchietti, L., Giardina, B., & Messana, I. (1996). Determination of peptide dissociation constants and stokes radius at different protonation stages by capillary electrophoresis. *Electrophoresis,17*(12), 1925-1930. doi:10.1002/elps.1150171220
44. Chambers, S. P., & Swalley, S. E. (2009).Designing Experiments for High-Throughput Protein Expression.*Methods in Molecular Biology High Throughput Protein Expression and Purification,* 19-29. doi:10.1007/978-1-59745-196-3_2
45. Chandler, D. L., & MIT News Office. (2015, June 19). Explained: Chemical vapor deposition. Retrieved May 21, 2020, from http://news.mit.edu/2015/explained-chemical-vapor-deposition-0619
46. Chapter 6.(n.d.). Retrieved May 20, 2020, from http://www.bioinfo.org.cn/book/biochemistry/chapt06/bio1.htm
47. Charlesworth, C., Grady, R. R., Shin, L., Vale, W. W., Rivier, C., Rivier, J., & Schwartz, N. B. (1984). Differential Suppression of FSH and LH Secretion by Follicular Fluid in the Presence or Absence of GnRH. *Neuroendocrinology,38*(3), 199-205. doi:10.1159/000123891
48. Chennamsetty, N., Helk, B., Voynov, V., Kayser, V., & Trout, B. L. (2009, 08). Aggregation-Prone Motifs in Human Immunoglobulin G. *Journal of Molecular Biology,391*(2), 404-413. doi:10.1016/j.jmb.2009.06.028

49.

50. Chen, H., England, L., & Kane, C. M. (1992, 07). Characterization of a HeLa cDNA clone encoding the human SII protein, an elongation factor for RNA polymerase II. *Gene,116*(2), 253-258. doi:10.1016/0378-1119(92)90522-q

51. Chen, H., Shaffer, P. L., Huang, X., & Rose, P. E. (2013, 03).Rapid screening of membrane protein expression in transiently transfected insect cells.*Protein Expression and Purification,88*(1), 134-142. doi:10.1016/j.pep.2012.12.003

52. Chen, L., Mathews, F. S., Davidson, V. L., Huizinga, E. G., Vellieux, F. M., & Hol, W. G. (1992, 10). Three-dimensional structure of the quinoprotein methylamine dehydrogenase fromParacoccus denitrificans determined by molecular replacement at 2.8 Å resolution. *Proteins: Structure, Function, and Genetics,14*(2), 288-299. doi:10.1002/prot.340140214

53. Chen, Y., & Hwang, J. (2005, 09). Prediction of disulfide connectivity from protein sequences. *Proteins: Structure, Function, and Bioinformatics,61*(3), 507-512. doi:10.1002/prot.20627

54. Cheng, X. (2004, 04). Multidimensional separation and hyphenated techniques in pharmaceutical research: Practical considerations. *Drug Discovery Today,9*(7), 305-306. doi:10.1016/s1359-6446(04)03024-7

55. Chiou, J. C., Hung, C., & Lin, C. (2010, 08). Design, fabrication, and package of MEMS-based image stabilizer for photographic cell phone applications. *2010 International Conference on Optical MEMS and Nanophotonics*. doi:10.1109/omems.2010.5672155

56. Chiu, H., Liao, H., Shao, Y., Lu, Y., Lin, C., Tsai, I., & Kuo, C. (2018, 08). Development of a general method for quantifying IgG-based therapeutic monoclonal antibodies in human plasma using protein G purification coupled with a two internal standard calibration strategy using LC-MS/MS. *Analytica Chimica Acta, 1019*, 93-102. doi:10.101/j.aca.2018.02.040

57. Chong, J., Hatsis, P., & Lucy, C. A. (2003, 05).High-speed ion chromatographic separation of cations at elevated temperature. *Journal of Chromatography A,997*(1-2), 161-169. doi:10.1016/s0021-9673(03)00549-1

58. Chrambach, A., Reisfeld, R., Wyckoff, M., & Zaccari, J. (1967, 07).A procedure for rapid and sensitive staining of protein fractionated by

polyacrylamide gel electrophoresis.*Analytical Biochemistry,20*(1), 150-154. doi:10.1016/0003-2697(67)90272-2

59. Clonis, Y. D. (2006, 01). Affinity chromatography matures as bioinformatic and combinatorial tools develop. *Journal of ChromatographyA,1101*(1-2),1-24.doi:10.1016/j.chroma.2005.09.073
60. Cohen, B. I., & Cohen, F. E. (1994).Predictions of Protein Secondary and Tertiary Structure.*Biocomputing,* 203-232. doi:10.1016/b978-0-08-092596-7.50010-1
61. Cort, B., Perkins, B., & Cort, G. (1995, 05).The NMT-5 criticality database.doi:10.2172/70707
62. Creative Proteomics.(n.d.).2D Blue Native/SDS-PAGE for Complex Analysis. Retrieved May 20, 2020, from https://www.creative-proteomics.com/services/2d-blue-native-sds-page-for-complex-analysis.htm
63. Cuatrecasas, P., & Wilchek, M. (2004).Affinity Chromatography. *Encyclopedia of Biological Chemistry,* 51-56. doi:10.1016/b0-12-443710-9/00011-9
64. Cuello, A., & Milstein, C. (1982).Monoclonal Antibodies and Neurotransmitters.*Neurotransmitters, Receptors,* 167-176. doi:10.1016/b978-0-08-028022-6.50020-5
65. Cummins, P. M., Rochfort, K. D., & O'Connor, B. F. (1970, January 01). Ion-Exchange Chromatography: Basic Principles and Application. Retrieved May 20, 2020, from https://link.springer.com/protocol/10.1007/978-1-4939-6412-3_11
66. Damen, C. W., Groot, E. R., Heij, M., Boss, D. S., Schellens, J. H., Rosing, H., . . . Aarden, L. A. (2009, 08). Development and validation of an enzyme-linked immunosorbent assay for the quantification of trastuzumab in human serum and plasma. *Analytical Biochemistry,391*(2), 114-120. doi:10.1016/j.ab.2009.05.030
67. Delaunay, J. (1995). Red Cell Membrane.*Molecular Basis of Human Blood Group Antigens Blood Cell Biochemistry,* 1-36. doi:10.1007/978-1-4757-9537-0_1
68. Destro-Bisol, G., Dobosz, M., & Pascali, V. (2000). ELECTROPHORESIS | Cellulose Acetate.*Encyclopedia of Separation Science,* 1222-1227. doi:10.1016/b0-12-226770-2/03541-9
69. Dhanda, S., Singh, H., Singh, J., & Singh, T. P. (2007, 04).Isolation, purification and characterization of a DPP-III homologue from

goat brain.*Protein Expression and Purification,52*(2), 297-305. doi:10.1016/j.pep.2006.10.004

70. Dutra, C., Pezo, D., Freire, M. T., Nerín, C., & Reyes, F. G. (2011, 03). Determination of volatile organic compounds in recycled polyethylene terephthalate and high-density polyethylene by headspace solid phase microextraction gas chromatography mass spectrometry to evaluate the efficiency of recycling processes.*Journal of Chromatography A,1218*(10), 1319-1330. doi:10.1016/j.chroma.2010.12.099
71. Edwards, H. (2006, 08).Spectra-Structure Correlations in Raman Spectroscopy.*Handbook of Vibrational Spectroscopy*. doi:10.1002/0470027320.s4103
72. Eng, J., Eng, J., Eng, J., & Eng, J. (2009, 07). Translational Proteomics 2.0 Meeting: SEQUEST and TPP Tips. *SciVee*. doi:10.4016/11920.01
73. Ettre, L. S., & Sakodynskii, K. I. (1993, 02). M. S. Tswett and the discovery of chromatography I: Early work (1899–1903). *Chromatographia,35*(3-4), 223-231. doi:10.1007/bf02269707
74. Ettre, L. S. (2001). Chapter 2 The builders of chromatography — Major chromatography awards and the award winners. *Journal of Chromatography Library Chromatography a Century of Discovery 1900–2000 - the Bridge to the Sciences/technology,* 39-54.doi:10.1016/s0301-4770(01)80008-7
75. Ewles, M., & Goodwin, L. (2011, 06). Bioanalytical approaches to analyzing peptides and proteins by LC–MS/MS. *Bioanalysis,3*(12), 1379-1397. doi:10.4155/bio.11.112
76. Evered, D. (1959, 11).Ionophoresis of acidic and basic amino acids on filter paper using low voltages.*Biochimica Et Biophysica Acta,36*(1), 14-19. doi:10.1016/0006-3002(59)90063-0
77. Fales, H. M. (1965). Some Practical Problems in Gas-Liquid Chromatography.*Gas Chromatography of Steroids in Biological Fluids,* 297-300. doi:10.1007/978-1-4684-0691-7_26
78. Fanali, C., Orazio, G. D., & Fanali, S. (2016). Chiral Separations Using Nano-Liquid Chromatography.*Scientia Chromatographica,8*(3), 161-169. doi:10.4322/sc.2016.029
79. Fekete, T. (2010).Bacillus Species and Related Genera Other than Bacillus anthracis.*Mandell, Douglas, and Bennett's Principles and Practice of Infectious Diseases,* 2727-2731. doi:10.1016/b978-0-443-06839-3.00209-5

80. Fernanda Homem De Mello De Souza Klein, Renata Homem De Mello De Souza Alves, & Andre Homem De Mello De Souza. (2013, 06). Case for diagnosis.*Anais Brasileiros De Dermatologia,88*(3), 465-467. doi:10.1590/abd1806-4841.20131764
81. Fitzpatrick, F., & Pike, J. (1981, 01). Factors affecting prostaglandin endoperoxide transformation: Antibodies against PGH2. *Progress in Lipid Research,20*, 179-182. doi:10.1016/0163-7827(81)90033-3
82. Flamini, R. (2013). Recent Applications of Mass Spectrometry in the Study of Grape and Wine Polyphenols.*ISRN Spectroscopy,2013*, 1-45. doi:10.1155/2013/813563
83. Fong, B. A., Wu, W., & Wood, D. W. (2009, 08). Optimization of ELP-intein mediated protein purification by salt substitution. *Protein Expression and Purification,66*(2), 198-202. doi:10.1016/j.pep.2009.03.009
84. Ford, A. E., & Morano, K. A. (2019). Thiol-Based Redox Signaling: Impacts on Molecular Chaperones and Cellular Proteostasis. *Heat Shock Proteins Heat Shock Proteins in Signaling Pathways,* 3-22. doi:10.1007/978-3-030-03952-3_1
85. Fourcroy, A. D., Lavoisier, A. L., Nicholson, W., & Sage, B. G. (1790). Elements of natural history and chemistry /. doi:10.5962/bhl.title.118578
86. Freeman, S. (2020, January 27). How DNA Profiling Works. Retrieved May 20, 2020, from https://science.howstuffworks.com/dna-profiling.htm
87. Freitag, R., & Horváth, C. (1995).Chromatography in the downstream processing of biotechnological products.*Downstream Processing Biosurfactants Carotenoids Advances in Biochemical Engineering/Biotechnology,* 17-59.doi:10.1007/bfb0102324
88. Gaca, J. G., Lee, W., Aksoy, O., Braedehoeft, S. J., Gonzalez-Stawinski, G. V., Parker, W., & Davis, R. (2001, 10). Evidence for polyreactive xenoreactive antibodies in the repertoire of human anti-swine antibodies: The 'next' humoral barrier to xenotransplantation? *Transplant Immunology,9*(1), 19-27. doi:10.1016/s0966-3274(01)00047-8
89. Gas Chromatography.(n.d.). Retrieved May 20, 2020, from https://www.sciencedirect.com/topics/agricultural-and-biological-sciences/gas-chromatography
90. Gel Electrophoresis. (n.d.). Retrieved May 20, 2020, from https://

www.sciencedirect.com/topics/neuroscience/gel-electrophoresis

91. Gelfi, C., & Righetti, P. G. (1981, 02).Polymerization kinetics of polyacrylamide gels I. Effect of different cross-linkers. *Electrophoresis,2*(4), 213-219. doi:10.1002/elps.1150020404
92. Gimba, E., & Tilli, T. (2013, 04). Human osteopontin splicing isoforms: Known roles, potential clinical applications and activated signaling pathways. *Cancer Letters,331*(1), 11-17. doi:10.1016/j.canlet.2012.12.003
93. Gleria, K. D., Hill, H., & Wong, L. (1996, 07). N -(2-Ferrocene-ethyl)maleimide : A new electroactive sulphydryl-specific reagent for cysteine-containing peptides and proteins. *FEBS Letters,390*(2), 142-144. doi:10.1016/0014-5793(96)00648-5
94. Goodsell, D. (2013, 02).Proton-Gated Urea Channel.*RCSB Protein Data Bank*. doi:10.2210/rcsb_pdb/mom_2013_2
95. Gottschalk, I., Lagerquist, C., Zuo, S., Lundqvist, A., & Lundahl, P. (2002, 02).Immobilized-biomembrane affinity chromatography for binding studies of membrane proteins.*Journal of Chromatography B,768*(1), 31-40. doi:10.1016/s0378-4347(01)00483-2
96. Gratz, A. J., Hillner, P. E., Manne, S., & Hansma, P. K. (1992, 05). Dynamic studies of crystal growth using the AFM.*Scanning Probe Microscopies*. doi:10.1117/12.58185
97. Guether, R., & Reddington, M. V. (1997, 09). Photostable Cyanine Dye β-Cyclodextrin Conjugates. *Tetrahedron Letters,38*(35), 6167-6170. doi:10.1016/s0040-4039(97)01426-3
98. Guillarme, D., & Veuthey, J. (2018, 07). "HPLC Teaching Simulator": A Simple Excel Tool for Teaching Liquid Chromatography. *Software-Assisted Method Development in High Performance Liquid Chromatography,* 187-216. doi:10.1142/9781786345462_0007
99. Gupta, S., & Tyagi, A. K. (1993, 04).Sequence of a newly identified Mycobacterium tuberculosis gene encoding a protein with sequence homology to virulence-regulating proteins.*Gene,126*(1), 157-158. doi:10.1016/0378-1119(93)90607-5
100. Hage, D. S. (2018).Chromatography.*Principles and Applications of Clinical Mass Spectrometry,* 1-32. doi:10.1016/b978-0-12-816063-3.00001-3
101. Hais, I. M. (1986, 04). Biospecific sorption, Prague, 1910: Emil Starkenstein (1884–1942). *Journal of Chromatography B:*

Biomedical Sciences and Applications,376, 5-9. doi:10.1016/s0378-4347(00)80820-8

102. Hais, I. M. (1986, 04). Biospecific sorption, Prague, 1910: Emil Starkenstein (1884–1942). *Journal of Chromatography B: Biomedical Sciences and Applications,376*, 5-9. doi:10.1016/s0378-4347(00)80820-8
103. Harrison, M., Langley, G., & Baldwin, M. (1989, 07). Effects of repellar position and voltage in thermospay mass spectrometry. *Journal of Chromatography A,474*(1), 139-143. doi:10.1016/s0021-9673(01)93908-1
104. Harwig, S., Chen, N., Park, A., & Lehrer, R. (1993, 02).Purification of Cysteine-Rich Bioactive Peptides from Leukocytes by Continuous Acid-Urea-Polyacrylamide Gel Electrophoresis.*Analytical Biochemistry,208*(2), 382-386. doi:10.1006/abio.1993.1065
105. Heger, N. E., & Kellogg, M. D. (2014, 05). Personalized Therapeutics and Companion Diagnostics: A New Paradigm in Testing and Treatment. *Clinical Chemistry,60*(5), 795-796. doi:10.1373/clinchem.2013.220228
106. He, F. (2011).Laemmli-SDS-PAGE.*Bio-Protocol,1*(11). doi:10.21769/bioprotoc.80
107. He, Q., Fang, Q., Du, W., & Fang, Z. (2007, 08).Fabrication of a monolithic sampling probe system for automated and continuous sample introduction in microchip-based CE.*Electrophoresis,28*(16), 2912-2919. doi:10.1002/elps.200600611
108. Hinson, G., & Williams, P. (1989).Adhesins of Pathogenic Escherichia coli.*Genetics of Bacterial Diversity,* 287-307. doi:10.1016/b978-0-12-355574-8.50023-9
109. Hjertén,S.,&Mosbach,R.(1962,02)."Molecular-sieve"chromatography of proteins on columns of cross-linked polyacrylamide.*Analytical Biochemistry,3*(2), 109-118. doi:10.1016/0003-2697(62)90100-8
110. Ho, J., & Todorova, L. (2009, 05). Pgi5 Published Cost-Effectiveness Results For Chronic Hepatitis B And C-A Comparative Analysis. *Value in Health,12*(3). doi:10.1016/s1098-3015(10)73352-0
111. Ho, T., Ho, T., Fieser, M., Fieser, L., Danheiser, R., & Roush, W. (2006, 12). N-Benzylhydroxylamine.*Fieser and Fieser's Reagents for Organic Synthesis*. doi:10.1002/9780471264194.fos00833
112. Hodges, R., & Mant, C. (2002, 01).Analytical HPLC of Peptides.

Hplc Of Biological Macro- Molecules, Revised And Expanded. doi:10.1201/9780203911594.pt2

113. Hopfgartner, G., Henion, J. D., Piguet, C., & Williams, A. F. (1993, 06).Application of Electrospray Mass Spectrometry for Characterizing Supramolecular Coordination Complexes.*Helvetica Chimica Acta,76*(4), 1759-1766. doi:10.1002/hlca.19930760432
114. Hsu, S. M., & Raine, L. (1981, 11).Protein A, avidin, and biotin in immunohistochemistry.*Journal of Histochemistry & Cytochemistry,29*(11), 1349-1353. doi:10.1177/29.11.6172466
115. Http://ars.els-cdn.com/content/image/1-s2.0-S000632079900088. (n.d.). Retrieved May 20, 2020, from https://prezi.com/j03qeryjvphs/httparsels-cdncomcontentimage1-s20-s000632079900088/
116. Http://ars.els-cdn.com/content/image/1-s2.0-S000632079900088. (n.d.). Retrieved May 20, 2020, from https://prezi.com/j03qeryjvphs/httparsels-cdncomcontentimage1-s20-s000632079900088/
117. Huang, S. (2003, 11).Size Exclusion/ Hydrodynamic Chromatography. *Handbook Of Size Exclusion Chromatography And Related Techniques.* doi:10.1201/9780203913321.ch23
118. Huang, Z. (2011, 02).Optimization of a HPLC determination method for Evodia rutaecarpa.*China Journal of Chinese Materia Medica.* doi:10.4268/cjcmm20110422
119. Hudson, G. A., Haddow, J. E., & Ritchie, R. F. (1983, 05). Disparity in antiserum avidity as measured by nephelometry, radial immunodiffusion, and column-binding strength.*Clinica Chimica Acta,130*(2), 239-244. doi:10.1016/0009-8981(83)90121-3
120. Huff, T., Müller, C., & Hannappel, E. (1997, 10). HPLC and post-column derivatization with fluorescamine: Isolation of actin-sequestering β-thymosins by reversed-phase HPLC. *Analytica Chimica Acta,352*(1-3), 239-248. doi:10.1016/s0003-2670(97)00133-5
121. Huston, W. M., Harhangi, H. R., Leech, A. P., Butler, C. S., Jetten, M. S., Camp, H. J., & Moir, J. W. (2007, 01). Expression and characterisation of a major c-type cytochrome encoded by gene kustc0563 from Kuenenia stuttgartiensis as a recombinant protein in Escherichia coli. *Protein Expression and Purification,51*(1), 28-33. doi:10.1016/j.pep.2006.06.026
122. Hutchens, T., Nelson, R. W., Li, C. M., & Yip, T. (1992, 06).Synthetic metal-binding protein surface domains for metal ion-dependent

interaction chromatography.*Journal of Chromatography A,604*(1), 125-132. doi:10.1016/0021-9673(92)85537-4

123. Hutcheson, P., & Slavin, R. (2003, 02). Correlation of serum precipitins and IgG anti aspergillus fumigatus (IgG-Af) determined by fluoroenzymeimmunoassay. *Journal of Allergy and Clinical Immunology,111*(2). doi:10.1016/s0091-6749(03)81058-8
124. ISO 14644-1:1999. (2015, December 09). Retrieved May 20, 2020, from https://www.iso.org/standard/25052.html
125. Ingram, J. T. (1958, 02).Carcinoma Erysipelatodes and Carcinoma Telangiectaticum.*Archives of Dermatology,77*(2), 227. doi:10.1001/archderm.1958.01560020081011
126. Insepov, Z., Wolf, D., & Hassanein, A. (2006, 09).Nanopumping Using Carbon Nanotubes.*Nano Letters,6*(9), 1893-1895. doi:10.1021/nl060932m
127. Instructables. (2017, November 05). Fraction Collector for Chromatography (Lego Wedo and. Retrieved May 20, 2020, from https://www.instructables.com/id/Fraction-Collector-for-Chromatography-Lego-Wedo-a/
128. Introduction to Affinity Chromatography.(n.d.). Retrieved May 20, 2020, from https://www.bio-rad.com/en-us/applications-technologies/introduction-affinity-chromatography?ID=MWHAVG4VY
129. Ion Exchange.(n.d.). Retrieved May 20, 2020, from https://www.sciencedirect.com/topics/immunology-and-microbiology/ion-exchange
130. Ionization. (2020, April 12). Retrieved May 20, 2020, from https://en.wikipedia.org/wiki/Matrix-assisted_laser_desorption/ionization
131. Isoelectric Focusing in 2D Electrophoresis.(n.d.). Retrieved May 20, 2020, from https://www.bio-rad.com/en-us/applications-technologies/isoelectric-focusing-2d-electrophoresis?ID=LUSQLK2B7
132. Jacobson, M. R. (1982, 04). Chemical Vapor Deposition: Applying The Technology To Optics. *Optical Thin Films*. doi:10.1117/12.933295
133. Johansson, J., Johansson, T., & Nilsson, S. (1998, 09).Fluorescence imaging of light absorption for axial-beam geometry in capillary electrophoresis.*Electrophoresis,19*(12), 2233-2238. doi:10.1002/elps.1150191231
134. Jorgenson, J., & Lukacs, K. (1983, 10).Capillary zone electrophoresis. *Science,222*(4621), 266-272. doi:10.1126/science.6623076

135. Jovin, T., Chrambach, A., & Naughton, M. (1964, 11).An apparatus for preparative temperature-regulated polyacrylamide gel electrophoresis. *Analytical Biochemistry,9*(3), 351-369. doi:10.1016/0003-2697(64)90192-7

136. Kalyanpur, M. (2002).Downstream Processing in the Biotechnology Industry.*Molecular Biotechnology,22*(1), 087-098. doi:10.1385/mb:22:1:087

137. Kang, T., Han, S., Kim, M., Kim, Y., & Yang, M. (2004, 11). Expression of non-toxic mutant of Escherichia coli heat-labile enterotoxin in tobacco chloroplasts.*Protein Expression and Purification,38*(1), 123-128. doi:10.1016/j.pep.2004.08.002

138. Karger, B. L. (1967, 07).New Developments in Chemical Selectivity in Gas-Liquid Chromatography.*Analytical Chemistry,39*(8). doi:10.1021/ac60252a719

139. Karlsson, A., & Karlsson, O. (2001, 01).Chiral ion-pair chromatography on porous graphitized carbon using N-blocked dipeptides as counter ions.*Journal of Chromatography A,905*(1-2), 329-335. doi:10.1016/s0021-9673(00)00986-9

140. Kato, M., Li, J., Chuang, J., & Chuang, D. (2007, 07).Structure of pyruvate dehydrogenase kinase isoform 1. doi:10.2210/pdb2q8f/pdb

141. Keβler, L. C., Gueorguieva, L., Rinas, U., & Seidel-Morgenstern, A. (2007, 12). Step gradients in 3-zone simulated moving bed chromatography. *Journal of Chromatography A,1176*(1-2), 69-78. doi:10.1016/j.chroma.2007.10.087

142. Kelley, B. (2009, 09). Industrialization of mAb production technology: The bioprocessing industry at a crossroads. *MAbs,1*(5), 443-452. doi:10.4161/mabs.1.5.9448

143. Kendrew, J. C., & Perutz, M. F. (1957, 06).X-Ray Studies of Compounds of Biological Interest.*Annual Review of Biochemistry,26*(1), 327-372. doi:10.1146/annurev.bi.26.070157.001551

144. Keren, D. F. (1994). Conditions Associated With Monoclonal Gammopathies. *High-Resolution Electrophoresis and Immunofixation,* 211-282. doi:10.1016/b978-0-7506-9469-8.50012-1

145. Khandurina, J., & Guttman, A. (2002, 01).Bioanalysis in microfluidic devices.*Journal of Chromatography A,943*(2), 159-183. doi:10.1016/s0021-9673(01)01451-0

146. Klobusitzky, D. V. (1939, 07).Electro-ultrafilter for Industrial Use.

The Journal of Physical Chemistry,43(7), 823-824. doi:10.1021/j150394a001

147. Kohn, J. (1957, 08).A cellulose acetate supporting medium for zone electrophoresis.*Clinica Chimica Acta,2*(4), 297-303. doi:10.1016/0009-8981(57)90005-0
148. Kondo, T., Mukai, M., & Kondo, Y. (1991, 10).Rapid isolation of plasmid DNA by LiCl-ethidium bromide treatment and gel filtration.*Analytical Biochemistry,198*(1), 30-35. doi:10.1016/0003-2697(91)90501-j
149. Kovacs, A., & Mescheder, U. M. (2001). Surface Micromachining Process for C-Si as Active Material. *Transducers '01 Eurosensors XV,* 620-623. doi:10.1007/978-3-642-59497-7_147
150. Kroeff, E. P. (2020, 02).Chromatographic Purification of Recombinant Protein Products.*Separations Technology,* 271-301. doi:10.1201/9780367813321-6
151. Kuehnbaum, N. L., Kormendi, A., & Britz-Mckibbin, P. (2013, 11). Multisegment Injection-Capillary Electrophoresis-Mass Spectrometry: A High-Throughput Platform for Metabolomics with High Data Fidelity. *Analytical Chemistry,85*(22), 10664-10669. doi:10.1021/ac403171u
152. Kumaresan, R., & Kolandaivel, P. (1995, 01).Structure and Stability of the Formic Acid Monomer and the Formic Acid—Water Complex. Anab initioStudy.*Zeitschrift Für Physikalische Chemie,192*(Part_2), 191-199. doi:10.1524/zpch.1995.192.part_2.191
153. Labrou, N. (2003, 06).Design and selection of ligands for affinity chromatography.*Journal of Chromatography B,790*(1-2), 67-78. doi:10.1016/s1570-0232(03)00098-9
154. Lacki, K. M., & Brekkan, E. (2011, 03).High Throughput Screening Techniques in Protein Purification.*Methods of Biochemical Analysis Protein Purification,* 487-506. doi:10.1002/9780470939932.ch20
155. Larson, S. M. (1994, 08). Clinical Studies Of Radioimmunodetection. *Journal of Immunotherapy,16*(2), 151. doi:10.1097/00002371-199408000-00021
156. Lee, H. (2004, 09).Microchip capillary electrophoresis with amperometric detection for several carbohydrates.*Talanta,64*(1), 210-216. doi:10.1016/j.talanta.2004.02.013
157. Licklider, L. J., Thoreen, C. C., Peng, J., & Gygi, S. P. (2002, 07). Automation of Nanoscale Microcapillary Liquid Chromatography–

Tandem Mass Spectrometry with a Vented Column.*Analytical Chemistry,74*(13), 3076-3083. doi:10.1021/ac025529o

158. Listek, M., Hönow, A., Gossen, M., & Hanack, K. (2020, February 03). A novel selection strategy for antibody producing hybridoma cells based on a new transgenic fusion cell line. Retrieved May 20, 2020, from https://www.nature.com/articles/s41598-020-58571-w
159. Lookhart, G. L., Bean, S. R., & Jones, B. L. (1999, 06).Separation and characterization of barley (Hordeum vulgareL.) hordeins by free zone capillary electrophoresis.*Electrophoresis,20*(7), 1605-1612. doi:10.1002/(sici)1522-2683(19990601)20:73.0.co;2-j
160. Loomis, C. W., & Brien, J. F. (1981, 03).Determination of carbimide in plasma by gas—liquid chromatography.*Journal of Chromatography B: Biomedical Sciences and Applications,222*(3), 421-428. doi:10.1016/s0378-4347(00)84142-0
161. Lu, W. (1970, January 01). Fighting malnutrition one crop at a time :). Retrieved May 20, 2020, from https://project-a-maize-ing.blogspot.com/2014/02/reverse-phase-high-performance-liquid.html
162. Lundell, N., & Schreitmüller, T. (1999, 01).Sample Preparation for Peptide Mapping— A Pharmaceutical Quality-Control Perspective. *Analytical Biochemistry,266*(1), 31-47. doi:10.1006/abio.1998.2919
163. Lyubimova, T., & Righetti, P. G. (1993). On the kinetics of photopolymerization: A theoretical study. *Electrophoresis,14*(1), 191-201. doi:10.1002/elps.1150140132
164. Mackintosh, C., & Diplexcito, J. (2003).Naturally Occurring Inhibitors of Protein Serine/Threonine Phosphatases.*Handbook of Cell Signaling,* 607-611. doi:10.1016/b978-012124546-7/50463-0
165. Manosalvaali. (2018, February 13). Thin-Layer Chromatography : Manual Method Instrumentation Tools. Retrieved May 20, 2020, from https://instrumentationtools.com/thin-layer-chromatography-manual-method/
166. Manz, A., Fröbe, Z., & Simon, W. (1985). Femtoliter Cell Volume Potentiometric Detector For Open-Tubular Column Liquid Chromatography. *Journal of Chromatography Library,* 297-307. doi:10.1016/s0301-4770(08)60837-4
167. Margolis, J., & Kenrick, K. G. (1967, 06).Polyacrylamide Gel-electrophoresis across a Molecular Sieve Gradient.*Nature,214*(5095), 1334-1336. doi:10.1038/2141334a0

168. Marshall, T., Abbott, N. J., Fox, P., & Williams, K. M. (1995). Protein concentration by precipitation with pyrogallol red prior to electrophoresis.*Electrophoresis,16*(1), 28-31. doi:10.1002/elps.1150160107

169. Maurer, H. R. (1978, 01).Disc Electrophoresis and Related Techniques of Polyacrylamide Gel Electrophoresis.doi:10.1515/9783110836202

170. Maurer, H. R. (1978, 01).Disc Electrophoresis and Related Techniques of Polyacrylamide Gel Electrophoresis.doi:10.1515/9783110836202

171. May / June 2009. (n.d.). Retrieved May 20, 2020, from https://ispe.org/pharmaceutical-engineering/may-june-2009

172. Mccroskey, M. C., Groppi, V. E., & Pearson, J. D. (1987, 06). Separation and purification of S49 mouse lymphoma histones by reversed-phase high-performance liquid chromatography. *Analytical Biochemistry,163*(2), 427-432. doi:10.1016/0003-2697(87)90244-2

173. Medana, C., Calza, P., Baiocchi, C., & Pelizzetti, E. (2005, 06).Liquid Chromatography Tandem Mass Spectrometry as a Tool to Investigate Pesticides and Their Degradation Products.*Current Organic Chemistry,9*(9), 859-873. doi:10.2174/1385272054038255

174. Medie, F. M., Vincentelli, R., Drancourt, M., & Henrissat, B. (2011, 02). Mycobacterium tuberculosis Rv1090 and Rv1987 encode functional β-glucan-targeting proteins. *Protein Expression and Purification,75*(2), 172-176. doi:10.1016/j.pep.2010.08.015

175. Mellors, J. S. (2016, 11).Microchip Capillary Electrophoresis with Integrated Electrospray-Mass Spectrometry for the Analysis of Small Molecule Metabolites, Digested Proteins, and Intact Monoclonal Antibodies. doi:10.26226/morressier.57d6b2bdd462b8028d88e2b2

176. Mendez, G., Gonzalez, G., & Mendez, G. (1979, 11).Radiologic Appearance of Pulmonary South American Blastomycosis.*Southern Medical Journal,72*(11), 1399-1401. doi:10.1097/00007611-197911000-00016

177. Mhatre, M. (1999, 05).Increased Ro15-4513–Induced Seizures Following Multiple Ethanol Withdrawals.*Pharmacology Biochemistry and Behavior,63*(1), 93-99. doi:10.1016/s0091-3057(98)00257-3

178. Miller, J., & Mutzelberg, I. (1973, 01).Laurell electrophoresis on cellulose acetate.*Journal of Chromatography A,75*(1), 165-168. doi:10.1016/s0021-9673(01)83432-4

179. Minc, N., Bokov, P., Zeldovich, K. B., Fütterer, C., Viovy, J., &

Dorfman, K. D. (2005, 01).Motion of single long DNA molecules through arrays of magnetic columns.*Electrophoresis,26*(2), 362-375. doi:10.1002/elps.200410115

180. Miyabe, T., Iida, H., Ohnishi, A., & Yashima, E. (2012). Enantioseparation on poly(phenyl isocyanide)s with macromolecular helicity memory as chiral stationary phases for HPLC. *Chem. Sci.,3*(3), 863-867. doi:10.1039/c1sc00708d
181. Monroe, C. W., & Harris, R. G. (1965, 03).The silver staining of reticular fibers by electrophoresis.*The Anatomical Record,151*(3), 239-242. doi:10.1002/ar.1091510304
182. Muronetz, V. I., & Korpela, T. (2003, 06).Isolation of antigens and antibodies by affinity chromatography.*Journal of Chromatography B,790*(1-2), 53-66. doi:10.1016/s1570-0232(03)00131-4
183. Murphy, A., & Fágáin, C. Ó. (1997, 01).Dipeptide synthesis using acetylated trypsin derivative.*Biotechnology Techniques,11*(1), 13-16. doi:10.1007/bf02764443
184. Murphy, L. K., & Targan, S. R. (1996).Antineutrophil Cytoplasmic Antibodies in Inflammatory Bowel Diseases.*Autoantibodies,* 47-52. doi:10.1016/b978-044482383-0/50009-1
185. Myers, M. N., & Giddings, J. C. (1965, 11). High Column Efficiency in Gas Liquid Chromatography at Inlet Pressures to 2500 p.s.i. *Analytical Chemistry,37*(12), 1453-1457. doi:10.1021/ac60231a003
186. Neuhoff, V., Stamm, R., & Eibl, H. (1985).Quantitative Aspects of Protein Staining by Coomassie Blue Dyes and Silver Staining.*Protides of the Biological Fluids,* 587-588. doi:10.1016/b978-0-08-033215-4.50143-6
187. Nevedomskaya, E., Mayboroda, O. A., & Deelder, A. M. (2011). Cross-platform analysis of longitudinal data in metabolomics.*Molecular BioSystems,7*(12), 3214. doi:10.1039/c1mb05280b
188. Nisnevitch, M., & Firer, M. (2001, 10). The solid phase in affinity chromatography: Strategies for antibody attachment. *Journal of Biochemical and Biophysical Methods,49*(1-3), 467-480. doi:10.1016/s0165-022x(01)00214-7
189. Northrup, M., Christel, L., Mcmillan, W., Petersen, K., Pourahmadi, F., Western, L., & Young, S. (1999). A new generation of PCR instruments and nucleic acid concentration systems.*PCR Applications,* 105-125. doi:10.1016/b978-012372185-3/50009-2

190. Odense, P. H., & Leung, T. C. (1975). Isoelectric Focusing On Polyacrylamide Gel And Starch Gel Electrophoresis Of Some Gadiform Fish Lactate Dehydrogenase Isozymes. *Isozymes,* 485-501. doi:10.1016/b978-0-12-472703-8.50036-6

191. Osborn, A. (1968, 12).HE formulation.Quarterly report, January--March, 1968.doi:10.2172/527452

192. Paech, C., & Christianson, T. (1994).Zymography of Proteases.*Cell Biology,* 264-271. doi:10.1016/b978-0-12-164717-9.50038-6

193. Pall, G. S., & Hamilton, A. J. (2008, June 05). Improved northern blot method for enhanced detection of small RNA. Retrieved May 20, 2020, from https://www.nature.com/articles/nprot.2008.67

194. Palomo-Jiménez, P. I., Hernández-Hernando, S., García-Nieto, R. M., & Villalobo, A. (1999, 08). A Method for the Purification of Phospho(Tyr)calmodulin Free of Nonphosphorylated Calmodulin. *Protein Expression and Purification,16*(3), 388-395. doi:10.1006/prep.1999.1092

195. Park, B. Y., Zaouk, R., & Madou, M. (2004, 05). Validation of lithography based on the controlled movement of light-emitting particles. *Emerging Lithographic Technologies VIII.* doi:10.1117/12.534355

196. Park, C., Son, C. K., Park, J. H., & Chung, I. S. (1997, 12).Separation and purification of useful proteins using hydrogel ultrafiltration. *Biotechnology and Bioprocess Engineering,2*(2), 113-116. doi:10.1007/bf02932337

197. Patil, M. (2020, January 10). Paper Chromatography: Instrumentation. Retrieved May 20, 2020, from https://microbenotes.com/paper-chromatography/

198. Pearson, J. D. (2004). HPLC Separation of Peptides.*Encyclopedia of Biological Chemistry,* 398-403. doi:10.1016/b0-12-443710-9/00321-5

199. Pennings, S., Meersseman, G., & Bradbury, E. (1992). Effect of glycerol on the separation of nucleosomes and bent DNA in low ionic strength polyacrylamide gel electrophoresis.*Nucleic Acids Research,20*(24), 6667-6672. doi:10.1093/nar/20.24.6667

200. Pesek, J. J., Matyska, M. T., Sentellas, S., Galceran, M. T., Chiari, M., & Pirri, G. (2002, 09). Multimodal open-tubular capillary electrochromatographic analysis of amines and peptides.*Electrophoresis,23*(17), 2982-2989. doi:10.1002/1522-2683(200209)23:173.0.co;2-y

201. Pethig, R., & Smith, S. (2012, 10). Introductory Bioelectronics. doi:10.1002/9781118443293

202. Pirici, D., Mogoanta, L., Kumar-Singh, S., Pirici, I., Margaritescu, C., Simionescu, C., & Stanescu, R. (2009, 02). Antibody Elution Method for Multiple Immunohistochemistry on Primary Antibodies Raised in the Same Species and of the Same Subtype. *Journal of Histochemistry & Cytochemistry, 57*(6), 567-575. doi:10.1369/jhc.2009.953240

203. Pontillo, N., & Merlino, A. (2016, 03). X-ray structure of cisplatin-encapsulated horse spleen apoferritin. doi:10.2210/pdb5erj/pdb

204. Porter, J. E., & Schuster, G. B. (1986, 03). ChemInform Abstract: Thermolysis of 4-Methyl-4-(1-propenyl)malonyl Peroxide: Mechanistic Limits to Chemiluminiscence Efficiency. *Chemischer Informationsdienst,17*(10). doi:10.1002/chin.198610219

205. Poulik, M. D. (1957, 12). Starch Gel Electrophoresis in a Discontinuous System of Buffers. *Nature,180*(4600), 1477-1479. doi:10.1038/1801477a0

206. Pridham, J. (1964). Paper Electrophoresis Of Phenolic Compounds. *Methods in Polyphenol Chemistry,* 111-124. doi:10.1016/b978-0-08-010887-2.50014-1

207. Qi, L., & Ito, Y. (2007, 06).Immunoaffinity centrifugal precipitation chromatography.*Journal of Chromatography A,1151*(1-2), 121-125. doi:10.1016/j.chroma.2007.02.119

208. Ramautar, R. (2016). Capillary Electrophoresis–Mass Spectrometry for Clinical Metabolomics.*Advances in Clinical Chemistry,* 1-34. doi:10.1016/bs.acc.2015.12.002

209. Ramsby, M. L., Makowski, G. S., & Khairallah, E. A. (1994). Differential detergent fractionation of isolated hepatocytes: Biochemical, immunochemical and two-dimensional gel electrophoresis characterization of cytoskeletal and noncytoskeletal compartments. *Electrophoresis,15*(1), 265-277. doi:10.1002/elps.1150150146

210. Rao, R., Rao, Y., & Shah, A. (1999, 08).A reverse phase ion-pairing HPLC method for the stability monitoring of sulphacetamide ophthalmic preparations.*Journal of Pharmaceutical and Biomedical Analysis,20*(4), 717-722. doi:10.1016/s0731-7085(99)00012-6

211. Raymond, S., & Weintraub, L. (1959, 09).Acrylamide Gel as a Supporting Medium for Zone Electrophoresis.*Science,130*(3377), 711-711. doi:10.1126/science.130.3377.711

212. Riederer, I. M., Herrero, R. M., Leuba, G., & Riederer, B. M. (2008, 07). Serial protein labeling with infrared maleimide dyes to identify cysteine modifications. *Journal of Proteomics,71*(2), 222-230. doi:10.1016/j.jprot.2008.04.006

213. Righetti, P. (2005). ELECTROPHORESIS | Two-Dimensional Gels. *Encyclopedia of Analytical Science,* 407-419. doi:10.1016/b0-12-369397-7/00126-6

214. Righetti, P. G., Gelfi, C., & Gianazza, E. (1986).Conventional Isoelectric Focusing and Immobilised pH Gradients.*Gel Electrophoresis of Proteins,* 141-202. doi:10.1016/b978-0-7236-0882-0.50008-3

215. Rivier, C., Rivier, J., Perrin, M., & Vale, W. (1983, 09).Comparison of the Effect of Several Gonadotropin Releasing Hormone Antagonists on Luteinizing Hormone Secretion, Receptor Binding and Ovulation.*Biology of Reproduction,29*(2), 374-378. doi:10.1095/biolreprod29.2.374

216. Rodríguez, I., Jin, L. J., & Li, S. F. (2000, 01). High-speed chiral separations on microchip electrophoresis devices.*Electrophoresis,21*(1), 211-219. doi:10.1002/(sici)1522-2683(20000101)21:13.0.co;2-d

217. Rojas, E. E., Coimbra, J., Saraiva, S., & Vicente, A. (2011, 02). Modeling of the α-lactalbumin and β-lactoglobulin protein separation.*Chemical Engineering Research and Design,89*(2), 156-163. doi:10.1016/j.cherd.2010.06.002

218. Roque, A. C., Silva, C. S., & Taipa, M. Â. (2007, 08). Affinity-based methodologies and ligands for antibody purification: Advances and perspectives. *Journal of Chromatography A,1160*(1-2), 44-55. doi:10.1016/j.chroma.2007.05.109

219. Rosner, W., & Kahn, S. M. (2005, 04).Autocrine and Paracrine Control of Breast Cancer Growth by Sex Hormone-Binding Globulin. doi:10.21236/ada437693

220. Sasajima, Y., Lim, L. W., Takeuchi, T., Suenami, K., Sato, K., & Takekoshi, Y. (2010).Simultaneous Determination of Antidepressants by Non-aqueous or Quasi-non-aqueous Capillary Electrophoresis. *Analytical Sciences,26*(6), 693-698. doi:10.2116/analsci.26.693

221. Sauve, D., Ho, D., & Roberge, M. (1995, 04).Concentration of Dilute Protein for Gel Electrophoresis.*Analytical Biochemistry,226*(2), 382-383. doi:10.1006/abio.1995.1242

222. Savarino, A., Boelaert, J. R., Cassone, A., Majori, G., & Cauda, R.

(2003, October 23). Effects of chloroquine on viral infections: An old drug against today's diseases. Retrieved May 21, 2020, from https://www.sciencedirect.com/science/article/pii/S1473309903008065

223. Schein, C. H. (2010, 08). ChemInform Abstract: Controlling Oligomerization of Pharmaceutical Proteins. *ChemInform,26*(21). doi:10.1002/chin.199521304
224. Schägger, H., & Jagow, G. V. (1991, 12).Blue native electrophoresis for isolation of membrane protein complexes in enzymatically active form.*Analytical Biochemistry,199*(2), 223-231. doi:10.1016/0003-2697(91)90094-a
225. Schmidt, C. L. (1943, 08).Proteins, Amino Acids and Peptides as Ions and Dipolar Ions (Cohn, Edwin J.; Edsall, John T.).*Journal of Chemical Education,20*(8), 415. doi:10.1021/ed020p415.2
226. Schoner, B. E., Belagaje, R. M., & Schoner, R. G. (1987). [25] Expression of eukaryotic genes in Escherichia coli with a synthetic two-cistron system. *Methods in Enzymology Recombinant DNA Part D,* 401-416. doi:10.1016/0076-6879(87)53068-3
227. Schüssler, W., & Nitschke, L. (1998, 05). Determination of trace amounts of Galaxolide ® (HHCB) by HPLC. *Fresenius' Journal of Analytical Chemistry,361*(2), 220-221. doi:10.1007/s002160050867
228. Schulte, I., Tammen, H., Selle, H., Zucht, H., & Schulz-Knappe, P. (2005). Chapter 7 Clinical peptidomics: Peptide-biomarker discovery in blood. *Proteomics and Peptidomics: New Technology Platforms Elucidating Biology Comprehensive Analytical Chemistry,* 385-409. doi:10.1016/s0166-526x(05)46007-9
229. Søeborg, T., Rasmussen, C., Mosekilde, E., & Colding-Jørgensen, M. (2012, 11).Biosimulation of Peptides and Proteins.*Pharmaceutical Formulation Development of Peptides and Proteins, Second Edition,* 323-338. doi:10.1201/b12951-14
230. Shapiro, H., Miller, K., & Harris, A. (1967, 12).Low-pH disc electrophoresis of spinal fluid; changes in multiple sclerosis. *Experimental and Molecular Pathology,7*(3), 362-365. doi:10.1016/0014-4800(67)90047-0
231. Shendure, J., & Aiden, E. L. (2012, 11).The expanding scope of DNA sequencing.*Nature Biotechnology,30*(11), 1084-1094. doi:10.1038/nbt.2421
232. Smithies, O. (1955, 12). Zone electrophoresis in starch gels: Group

variations in the serum proteins of normal human adults. *Biochemical Journal,61*(4), 629-641. doi:10.1042/bj0610629

233. Solignac, D., Sayah, A., Heimgartner, S., Szekely, L., Freitag, R., & Gijs, M. A. (2001).Electrochromatography On-Chip with Low Dead-Volume Lateral Capillary Connections.*Micro Total Analysis Systems 2001,* 597-598. doi:10.1007/978-94-010-1015-3_261
234. Song, S. M., & Kim, I. H. (2012, 08). Separation Characteristics of IgY (Immunoglobulin Yolk) in Various HPLC Columns. *Korean Chemical Engineering Research,50*(4), 659-665. doi:10.9713/kcer.2012.50.4.659
235. Stamets, P. E., Naeger, N. L., Evans, J. D., Han, J. O., Hopkins, B. K., Lopez, D., . . . Sheppard, W. S. (2018, October 04). Extracts of Polypore Mushroom Mycelia Reduce Viruses in Honey Bees. Retrieved May 20, 2020, from https://www.nature.com/articles/s41598-018-32194-8
236. Srbek, J., Eickhoff, J., Effelsberg, U., Kraiczek, K., Goor, T. V., & Coufal, P. (2007, 08). Chip-based nano-LC-MS/MS identification of proteins in complex biological samples using a novel polymer microfluidic device. *Journal of Separation Science, 30*(13), 2046-2052. doi:10.1002/jssc.200700053
237. Steinberg, T. H. (2009). Chapter 31 Protein Gel Staining Methods. *Methods in Enzymology Guide to Protein Purification, 2nd Edition,* 541-563. doi:10.1016/s0076-6879(09)63031-7
238. Strandberg, B. (2009, 09). Chapter 1: Building the Ground for the First Two Protein Structures: Myoglobin and Haemoglobin. *Journal of Molecular Biology,392*(1), 2-10. doi:10.1016/j.jmb.2009.05.087
239. Sun, X. (2004).Multitasking of Helix-Loop-Helix Proteins in Lymphopoiesis.*Advances in Immunology,* 43-77. doi:10.1016/s0065-2776(04)84002-1
240. Sussman, L. K. (1988). The Use of Herbal and Biomedical Pharmaceuticals on Mauritius.*The Context of Medicines in Developing Countries,* 199-215. doi:10.1007/978-94-009-2713-1_10
241. Tanaka, F., & Barbas, C. F. (2005, 05). Reactive Immunization: A Unique Approach to Aldolase Antibodies. *Catalytic Antibodies,* 304-335. doi:10.1002/3527603662.ch11
242. Tanford, C. (1955, 12).Ionography.Electrophoresis in Stabilized Media.*Journal of the American Chemical Society,77*(24), 6719-6720. doi:10.1021/ja01629a139
243. Tanford, C. (1970). Protein Denaturation.*Advances in Protein Chemistry*

Advances in Protein Chemistry Volume 24, 1-95. doi:10.1016/s0065-3233(08)60241-7

244. Tempst, P., Woo, D. D., Templow, D. B., Aebersold, R., Hood, L. E., & Kent, S. B. (1986, 01). Microscale structure analysis of a high-molecular-weight, hydrophobic membrane glycoprotein fraction with platelet-derived growth factor-dependent kinase activity.*Journal of Chromatography A,359*, 403-412. doi:10.1016/0021-9673(86)80094-2
245. Terlecka, E. (2005, 08). Arsenic Speciation Analysis in Water Samples: A Review of The Hyphenated Techniques. *Environmental Monitoring and Assessment,107*(1-3), 259-284. doi:10.1007/s10661-005-3109-z
246. Teshima, G., & Canova-Davis, E. (1992, 11).Separation of oxidized human growth hormone variants by revesed-phase high-performance liquid chromatography.*Journal of Chromatography A,625*(2), 207-215. doi:10.1016/0021-9673(92)85204-7
247. Thakur, K. G., Jaiswal, R. K., Shukla, J. K., Praveena, T., & Gopal, B. (2010, 12). Over-expression and purification strategies for recombinant multi-protein oligomers: A case study of Mycobacterium tuberculosis σ/anti-σ factor protein complexes. *Protein Expression and Purification,74*(2), 223-230. doi:10.1016/j.pep.2010.06.018
248. The Editors of Encyclopaedia Britannica. (2018, March 26). Electroplating. Retrieved May 21, 2020, from https://www.britannica.com/technology/electroplating
249. The Editors of Encyclopaedia Britannica. (2020, February 06). Antibody. Retrieved May 20, 2020, from https://www.britannica.com/science/antibody
250. The Editors of Encyclopaedia Britannica. (2020, February 06). Antibody. Retrieved May 20, 2020, from https://www.britannica.com/science/antibody
251. The Protein Protocols Handbook: John M. Walker. (n.d.). Retrieved May 20, 2020, from https://www.springer.com/gp/book/9781603272599
252. The band 3 proteins: Anion transporters, binding proteins and senescent antigens. (1992). *Progress in Cell Research*.doi:10.1016/c2009-0-00504-2
253. The principle and method of chromatography.(n.d.). Retrieved May 20, 2020, from https://www.mblbio.com/bio/g/support/method/chromatography.html
254. Thomsson, K. A., Bäckström, M., Larsson, J. M., Hansson, G. C., &

Karlsson, H. (2010, 02). Enhanced Detection of Sialylated and Sulfated Glycans with Negative Ion Mode Nanoliquid Chromatography/Mass Spectrometry at High pH. *Analytical Chemistry,82*(4), 1470-1477. doi:10.1021/ac902602e

255. Tiselius, A., Henschen, G. E., & Svensson, H. (1938, 10). Electrophoresis of pepsin. *Biochemical Journal,32*(10), 1814-1818. doi:10.1042/bj0321814
256. Tolosa, J., Graziani, G., Gaspari, A., Chianese, D., Ferrer, E., Mañes, J., & Ritieni, A. (2017, 02). Multi-Mycotoxin Analysis in Durum Wheat Pasta by Liquid Chromatography Coupled to Quadrupole Orbitrap Mass Spectrometry. *Toxins,9*(2), 59. doi:10.3390/toxins9020059
257. Tulchin, N., Ornstein, L., & Davis, B. J. (1976, 05). A microgel system for disc electrophoresis. *Analytical Biochemistry,72*(1-2), 485-490. doi:10.1016/0003-2697(76)90558-3
258. US4667016A - Erythropoietin purification.(n.d.). Retrieved May 20, 2020, from https://patents.google.com/patent/US4667016A/en
259. Ueda, M., Kiba, Y., Abe, H., Arai, A., Nakanishi, H., & Baba, Y. (2000, 01). Fast separation of oligonucleotide and triplet repeat DNA on a microfabricated capillary electrophoresis device and capillary electrophoresis. *Electrophoresis,21*(1), 176-180. doi:10.1002/(sici)1522-2683(20000101)21:13.0.co;2-a
260. Ujiie, T., Kikuchi, T., Ichiki, T., & Horiike, Y. (2000, 06). Fabrication of Quartz Microcapillary Electrophoresis Chips Using Plasma Etching. *Japanese Journal of Applied Physics,39*(Part 1, No. 6A), 3677-3682. doi:10.1143/jjap.39.3677
261. Ultrafiltration.(n.d.). Retrieved May 20, 2020, from https://www.sciencedirect.com/topics/medicine-and-dentistry/ultrafiltration
262. Ultrasonication.(n.d.). Retrieved May 20, 2020, from https://www.sciencedirect.com/topics/engineering/ultrasonication
263. Valdivia-Arenas, M. A., Soriano, A. O., & Arteaga, R. B. (2007). Constrictive Bronchiolitis after Treatment of Colon Cancer with 5-Fluorouracil. *Chemotherapy,53*(5), 316-319. doi:10.1159/000107456
264. Vilgalys, R., & Gonzalez, D. (1990, 10). Organization of ribosomal DNA in the basidiomycete Thanatephorus praticola. *Current Genetics,18*(3), 277-280. doi:10.1007/bf00318394
265. Vrecko, K., Staedtler, P., Mischak, I., Maresch, L., & Reibnegger, G. (1997, 12). Periodontitis and concentrations of the cellular immune

activation marker neopterin in saliva and urine.*Clinica Chimica Acta,268*(1-2), 31-40. doi:10.1016/s0009-8981(97)00154-x

266. Walsh, G. (2010, 09). Biopharmaceutical benchmarks 2010. *Nature Biotechnology,28*(9), 917-924. doi:10.1038/nbt0910-917

267. Wang, H., Zhao, Z., & Guo, Y. (2012). Chemical and Biochemical Applications of MALDI TOF-MS Based on Analyzing the Small Organic Compounds. *Topics in Current Chemistry Applications of MALDI-TOF Spectroscopy,* 165-192. doi:10.1007/128_2012_364

268. Wasinger, V. C., Cordwell, S. J., Cerpa-Poljak, A., Yan, J. X., Gooley, A. A., Wilkins, M. R., . . . Humphery-Smith, I. (1995). Progress with gene-product mapping of the Mollicutes:Mycoplasma genitalium. *Electrophoresis,16*(1), 1090-1094. doi:10.1002/elps.11501601185

269. Waters.(n.d.). How Does High Performance Liquid Chromatography Work? Retrieved May 20, 2020, from https://www.waters.com/waters/en_US/How-Does-High-Performance-Liquid-Chromatography-Work?/nav.htm?cid=10049055

270. Wei, J., Jiang, L., Lou, C., & Zhu, Y. (2018). Simultaneous determination of eight phthalate ester plasticizers in sports beverages using supercritical fluid chromatography. *Chinese Journal of Chromatography,36*(7), 678. doi:10.3724/sp.j.1123.2018.01014

271. Westermeier, R. (2005, 09). Gel Electrophoresis. *ELS.* doi:10.1038/npg.els.0005335

272. Western Blotting.(n.d.). Retrieved May 20, 2020, from https://www.sciencedirect.com/topics/medicine-and-dentistry/western-blotting

273. Wieme, R. (1966, 01).Improved agar support for electrophoresis of LDH isoenzymes.*Clinica Chimica Acta,13*(1), 138-139. doi:10.1016/0009-8981(66)90283-x

274. Wild, D. (2013).Immunoassay for Beginners.*The Immunoassay Handbook,* 7-10. doi:10.1016/b978-0-08-097037-0.00002-6

275. Wilken, L. R., & Nikolov, Z. L. (2011, 09).Downstream Processing of Transgenic Plant Systems: Protein Recovery and Purification Strategies.*Molecular Farming in Plants: Recent Advances and Future Prospects,* 217-257. doi:10.1007/978-94-007-2217-0_11

276. Wu, J. X., Wilsey, C. B., Taylor, L., & Schuurman, G. W. (n.d.). Projected avifaunal responses to climate change across the U.S. National Park System. Retrieved May 20, 2020, from https://journals.plos.org/plosone/article?id=10.1371/journal.pone.0190557

277. Wu, N., Medina, J. C., & Lee, M. L. (2000, 09). Fast gas chromatography: Packed column solvating gas chromatography versus open tubular column gas chromatography. *Journal of Chromatography A,892*(1-2), 3-13. doi:10.1016/s0021-9673(00)00152-7

278. Xia, Y., & Whitesides, G. M. (2010, 06). ChemInform Abstract: Soft Lithography. *ChemInform,29*(25). doi:10.1002/chin.199825359

279. X-ray lithography. (2020, May 04). Retrieved May 21, 2020, from https://en.wikipedia.org/wiki/X-ray_lithography

280. Yadav, T. (2015, October 12). Paper electrophoresis. Retrieved May 20, 2020, from https://www.slideshare.net/TapeshwarYadav1/paper-electrophoresis

281. Yang, C. (2012, 06).Nanoscale Line Segment Fabrication Using Super-Resolution Near-Field Photolithography.*Scanning,34*(5), 284-294. doi:10.1002/sca.21015

282. Yang, W. C., Welsh, J. P., Lee, J., Cooke, J. P., & Swartz, J. R. (2011, 11). Solubility partner IF2 Domain I enables high yield synthesis of transducible transcription factors in Escherichia coli. *Protein Expression and Purification,80*(1), 145-151. doi:10.1016/j.pep.2011.06.017

283. Young, C. L., Britton, Z. T., & Robinson, A. S. (2012, 01). Recombinant protein expression and purification: A comprehensive review of affinity tags and microbial applications. *Biotechnology Journal,7*(5), 620-634. doi:10.1002/biot.201100155

284. Yuan, J., Xu, Z., Cheng, S., & Feng, L. (2002, 08). The aggregation of polystyrene-b-poly(ethylene oxide)-b-polystyrene triblock copolymers in aqueous solution. *European Polymer Journal,38*(8), 1537-1546. doi:10.1016/s0014-3057(02)00025-3

285. Zayas, J. F. (1997). Emulsifying Properties of Proteins.*Functionality of Proteins in Food,* 134-227. doi:10.1007/978-3-642-59116-7_4

286. Zech, M., Hartley, M. R., & Bohnert, H. J. (1981, 09).Binding Sites of E. coli DNA-dependent RNA polymerase on spinach chloroplast DNA.*Current Genetics,4*(1), 37-46. doi:10.1007/bf00376784

287. Zhang, J., Liu, H., & Katta, V. (2009).Structural characterization of intact antibodies by high-resolution LTQ Orbitrap mass spectrometry. *Journal of Mass Spectrometry*. doi:10.1002/jms.1700

288. Zhang, Y., Zhang, Y., Wang, G., Chen, W., He, P., & Wang, Q. (2016). Synthesis and characterization of a multimode stationary phase: Congo red derivatized silica in nano-flow HPLC. *The Analyst,141*(3), 1083-

1090. doi:10.1039/c5an02021b

289. Ji, J. A., Zhang, B., Cheng, W., & Wang, Y. J. (2009, 12). Methionine, tryptophan, and histidine oxidation in a model protein, PTH: Mechanisms and stabilization. *Journal of Pharmaceutical Sciences,98*(12), 4485-4500. doi:10.1002/jps.21746

290. Korfmacher, W. (2011, 06). High-resolution mass spectrometry will dramatically change our drug-discovery bioanalysis procedures. *Bioanalysis,3*(11), 1169-1171. doi:10.4155/bio.11.98

291. Lange, V., Picotti, P., Domon, B., & Aebersold, R. (2008, 01). Selected reaction monitoring for quantitative proteomics: A tutorial. *Molecular Systems Biology,4*(1), 222. doi:10.1038/msb.2008.61

292. Leader, B., Baca, Q. J., & Golan, D. E. (2008, 01). Protein therapeutics: A summary and pharmacological classification. *Nature Reviews Drug Discovery,7*(1), 21-39. doi:10.1038/nrd2399

293. Lee, M. S., & Kerns, E. H. (1999).LC/MS applications in drug development.*Mass Spectrometry Reviews,18*(3-4), 187-279. doi:10.1002/(sici)1098-2787(1999)18:3/43.0.co;2-k

294. Lisitsa, A., Moshkovskii, S., Chernobrovkin, A., Ponomarenko, E., & Archakov, A. (2014, 01). Profiling proteoforms: Promising follow-up of proteomics for biomarker discovery. *Expert Review of Proteomics,11*(1), 121-129. doi:10.1586/14789450.2014.878652

295. Liu, X., Valentine, S. J., Plasencia, M. D., Trimpin, S., Naylor, S., & Clemmer, D. E. (2007, 07). Mapping the human plasma proteome by SCX-LC-IMS-MS. *Journal of the American Society for Mass Spectrometry,18*(7), 1249-1264. doi:10.1016/j.jasms.2007.04.012

296. Marshall, A. G., Hendrickson, C. L., & Jackson, G. S. (1998). Fourier transform ion cyclotron resonance mass spectrometry: A primer. *Mass Spectrometry Reviews,17*(1), 1-35. doi:10.1002/(sici)1098-2787(1998)17:13.0.co;2-k

297. Mason, E. A., & Mcdaniel, E. W. (1988, 10). Transport Properties of Ions in Gases. doi:10.1002/3527602852

298. Mclafferty, F. W., Breuker, K., Jin, M., Han, X., Infusini, G., Jiang, H., . . . Begley, T. P. (2007, 11). Top-down MS, a powerful complement to the high capabilities of proteolysis proteomics.*FEBS Journal,274*(24), 6256-6268. doi:10.1111/j.1742-4658.2007.06147.x

299. Omenn, G. S., Guan, Y., & Menon, R. (2014, 07). A new class of protein cancer biomarker candidates: Differentially expressed splice variants

of ERBB2 (HER2/neu) and ERBB1 (EGFR) in breast cancer cell lines. *Journal of Proteomics,107*, 103-112. doi:10.1016/j.jprot.2014.04.012

300. Pavlou, A. K., & Reichert, J. M. (2004, 12).Recombinant protein therapeutics—success rates, market trends and values to 2010.*Nature Biotechnology,22*(12), 1513-1519. doi:10.1038/nbt1204-1513
301. Rifai, N., Gillette, M. A., & Carr, S. A. (2006, 08). Protein biomarker discovery and validation: The long and uncertain path to clinical utility. *Nature Biotechnology,24*(8), 971-983. doi:10.1038/nbt1235
302. Scott, A. M., Wolchok, J. D., & Old, L. J. (2012, 03).Antibody therapy of cancer.*Nature Reviews Cancer,12*(4), 278-287. doi:10.1038/nrc3236
303. Snijder, J., & Heck, A. J. (2014, 06).Analytical Approaches for Size and Mass Analysis of Large Protein Assemblies.*Annual Review of Analytical Chemistry,7*(1), 43-64. doi:10.1146/annurev-anchem-071213-020015
304. Song, Y., Schowen, R. L., Borchardt, R. T., & Topp, E. M. (2001, 02). Effect of 'pH' on the rate of asparagine deamidation in polymeric formulations: 'pH'–rate profile. *Journal of Pharmaceutical Sciences,90*(2), 141-156. doi:10.1002/1520-6017(200102)90:23.0.co;2-y
305. Tavares, R., Scherer, N. D., Pauletti, B. A., Araújo, E., Folador, E. L., Espindola, G., . . . Passetti, F. (2014, 01). SpliceProt: A protein sequence repository of predicted human splice variants. *Proteomics,14*(2-3), 181-185. doi:10.1002/pmic.201300078
306. Valentine, S. J., Plasencia, M. D., Liu, X., Krishnan, M., Naylor, S., Udseth, H. R., . . . Clemmer, D. E. (2006, 11). Toward Plasma Proteome Profiling with Ion Mobility-Mass Spectrometry.*Journal of Proteome Research,5*(11), 2977-2984. doi:10.1021/pr060232i
307. Wang, H., Cao, H., & Wang, Y. (2010, 01).Quantification ofN2-Carboxymethyl-2′-deoxyguanosine in Calf Thymus DNA and Cultured Human Kidney Epithelial Cells by Capillary High-Performance Liquid Chromatography–Tandem Mass Spectrometry Coupled with Stable Isotope Dilution Method.*Chemical Research in Toxicology,23*(1), 74-81. doi:10.1021/tx900286c
308. Washburn, M. P., Wolters, D., & Yates, J. R. (2001, 03).Large-scale analysis of the yeast proteome by multidimensional protein identification technology.*Nature Biotechnology,19*(3), 242-247. doi:10.1038/85686
309. Wu, J., Puppala, D., Feng, X., Monetti, M., Lapworth, A. L., & Geoghegan, K. F. (2013, 11).Chemoproteomic Analysis of Intertissue and Interspecies Isoform Diversity of AMP-activated Protein Kinase

(AMPK). *Journal of Biological Chemistry, 288*(50), 35904-35912. doi:10.1074/jbc.m113.508747

310. Yassine, H., Borges, C. R., Schaab, M. R., Billheimer, D., Stump, C., Reaven, P., . . . Nelson, R. (2013, 07). Mass spectrometric immunoassay and MRM as targeted MS-based quantitative approaches in biomarker development: Potential applications to cardiovascular disease and diabetes. *PROTEOMICS - Clinical Applications, 7*(7-8), 528-540. doi:10.1002/prca.201200028
311. Zheng, J., Mehl, J., Zhu, Y., Xin, B., & Olah, T. (2014, 03). Application and challenges in using LC–MS assays for absolute quantitative analysis of therapeutic proteins in drug discovery. *Bioanalysis, 6*(6), 859-879. doi:10.4155/bio.14.36
312. Zhou, S., Song, Q., Tang, Y., & Naidong, W. (2005, 01). Critical Review of Development, Validation, and Transfer for High Throughput Bioanalytical LC-MS/MS Methods. *Current Pharmaceutical Analysis, 1*(1), 3-14. doi:10.2174/1573412052953346
313. Bagal, D., Valliere-Douglass, J. F., Balland, A., & Schnier, P. D. (2010, 08). Resolving Disulfide Structural Isoforms of IgG2 Monoclonal Antibodies by Ion Mobility Mass Spectrometry. *Analytical Chemistry, 82*(16), 6751-6755. doi:10.1021/ac1013139
314. Betz, S. F. (1993, 10). Disulfide bonds and the stability of globular proteins. *Protein Science, 2*(10), 1551-1558. doi:10.1002/pro.5560021002
315. Butterfield, A. M., Chain, J. S., Ackermann, B. L., & Konrad, R. J. (2010, 12). Comparison of assay formats for drug-tolerant immunogenicity testing. *Bioanalysis, 2*(12), 1961-1969. doi:10.4155/bio.10.136
316. Gorman, J. J., Wallis, T. P., & Pitt, J. J. (2002, 05). Protein disulfide bond determination by mass spectrometry. *Mass Spectrometry Reviews, 21*(3), 183-216. doi:10.1002/mas.10025
317. Gray, W. R. (1993, 10). Disulfide structures of highly bridged peptides: A new strategy for analysis. *Protein Science, 2*(10), 1732-1748. doi:10.1002/pro.5560021017
318. Han, X., Jin, M., Breuker, K., & Mclafferty, F. W. (2006, 10). Extending Top-Down Mass Spectrometry to Proteins with Masses Greater Than 200 Kilodaltons. *Science, 314*(5796), 109-112. doi:10.1126/science.1128868
319. Hwa, J., Reeves, P. J., Klein-Seetharaman, J., Davidson, F., & Khorana,

H. G. (1999, 03). Structure and function in rhodopsin: Further elucidation of the role of the intradiscal cysteines, Cys-110, -185, and -187, in rhodopsin folding and function. *Proceedings of the National Academy of Sciences,96*(5), 1932-1935. doi:10.1073/pnas.96.5.1932

320. Mallick, P., Boutz, D. R., Eisenberg, D., & Yeates, T. O. (2002, 07). Genomic evidence that the intracellular proteins of archaeal microbes contain disulfide bonds.*Proceedings of the National Academy of Sciences,99*(15), 9679-9684. doi:10.1073/pnas.142310499
321. Mamathambika, B. S., & Bardwell, J. C. (2008, 11).Disulfide-Linked Protein Folding Pathways.*Annual Review of Cell and Developmental Biology,24*(1), 211-235. doi:10.1146/annurev.cellbio.24.110707.175333
322. Murad, W., & Singh, R. (2013, 06). ${rm MS2DB} $: A Software for Determination of Disulfide Bonds Using Multi-Ion Analysis. *IEEE Transactions on NanoBioscience,12*(2), 69-71. doi:10.1109/tnb.2012.2212029
323. Torosantucci, R., Schöneich, C., & Jiskoot, W. (2013, 09). Oxidation of Therapeutic Proteins and Peptides: Structural and Biological Consequences. *Pharmaceutical Research,31*(3), 541-553. doi:10.1007/s11095-013-1199-9
324. Zhang, B., Yang, Y., Yuk, I., Pai, R., Mckay, P., Eigenbrot, C., . . . Francissen, K. C. (2008, 04). Unveiling a Glycation Hot Spot in a Recombinant Humanized Monoclonal Antibody. *Analytical Chemistry,80*(7), 2379-2390. doi:10.1021/ac701810q
325. Han, T. H., & Zhao, B. (2014, 07).Absorption, Distribution, Metabolism, and Excretion Considerations for the Development of Antibody-Drug Conjugates.*Drug Metabolism and Disposition,42*(11), 1914-1920. doi:10.1124/dmd.114.058586
326. Heudi, O., Barteau, S., Zimmer, D., Schmidt, J., Bill, K., Lehmann, N., . . . Kretz, O. (2008, 06). Towards Absolute Quantification of Therapeutic Monoclonal Antibody in Serum by LC−MS/MS Using Isotope-Labeled Antibody Standard and Protein Cleavage Isotope Dilution Mass Spectrometry.*Analytical Chemistry,80*(11), 4200-4207. doi:10.1021/ac800205s
327. Leader, B., Baca, Q. J., & Golan, D. E. (2008, 01). Protein therapeutics: A summary and pharmacological classification. *Nature Reviews Drug Discovery,7*(1), 21-39. doi:10.1038/nrd2399

328. Ouyang, Z., Furlong, M. T., Wu, S., Sleczka, B., Tamura, J., Wang, H., . . . Jemal, M. (2012, 01). Pellet digestion: A simple and efficient sample preparation technique for LC–MS/MS quantification of large therapeutic proteins in plasma. *Bioanalysis,4*(1), 17-28. doi:10.4155/bio.11.286

329. Suh, M., Clark, D. J., Parmer, P. P., Fleischmann, R. D., Peterson, S. N., & Pieper, R. (2010, 06). Using chemical derivatization and mass spectrometric analysis to characterize the post-translationally modified Staphylococcus aureus surface protein G. *Biochimica Et Biophysica Acta (BBA) - Proteins and Proteomics,1804*(6), 1394-1404. doi:10.1016/j.bbapap.2010.02.006

330. Whiteaker, J. R., Zhao, L., Zhang, H. Y., Feng, L., Piening, B. D., Anderson, L., & Paulovich, A. G. (2007, 03). Antibody-based enrichment of peptides on magnetic beads for mass-spectrometry-based quantification of serum biomarkers.*Analytical Biochemistry,362*(1), 44-54. doi:10.1016/j.ab.2006.12.023

INDEX

R

S

T